MIX
Papier aus verantwortungsvollen Quellen
Paper from responsible sources
FSC® C105338

R. Rajasree
V.S. Harikumar

Endomycorrhizal Association in Sesame

Effects on Growth and Nutrition

Anchor Academic
Publishing

Rajasree, R., Harikumar, V.S.: Endomycorrhizal Association in Sesame. Effects on Growth and Nutrition, Hamburg, Anchor Academic Publishing 2016

Buch-ISBN: 978-3-96067-074-2
PDF-eBook-ISBN: 978-3-96067-574-7
Druck/Herstellung: Anchor Academic Publishing, Hamburg, 2016

Bibliografische Information der Deutschen Nationalbibliothek:
Die Deutsche Nationalbibliothek verzeichnet diese Publikation in der Deutschen Nationalbibliografie; detaillierte bibliografische Daten sind im Internet über http://dnb.d-nb.de abrufbar.

Bibliographical Information of the German National Library:
The German National Library lists this publication in the German National Bibliography. Detailed bibliographic data can be found at: http://dnb.d-nb.de

All rights reserved. This publication may not be reproduced, stored in a retrieval system or transmitted, in any form or by any means, electronic, mechanical, photocopying, recording or otherwise, without the prior permission of the publishers.

Das Werk einschließlich aller seiner Teile ist urheberrechtlich geschützt. Jede Verwertung außerhalb der Grenzen des Urheberrechtsgesetzes ist ohne Zustimmung des Verlages unzulässig und strafbar. Dies gilt insbesondere für Vervielfältigungen, Übersetzungen, Mikroverfilmungen und die Einspeicherung und Bearbeitung in elektronischen Systemen.

Die Wiedergabe von Gebrauchsnamen, Handelsnamen, Warenbezeichnungen usw. in diesem Werk berechtigt auch ohne besondere Kennzeichnung nicht zu der Annahme, dass solche Namen im Sinne der Warenzeichen- und Markenschutz-Gesetzgebung als frei zu betrachten wären und daher von jedermann benutzt werden dürften.

Die Informationen in diesem Werk wurden mit Sorgfalt erarbeitet. Dennoch können Fehler nicht vollständig ausgeschlossen werden und die Diplomica Verlag GmbH, die Autoren oder Übersetzer übernehmen keine juristische Verantwortung oder irgendeine Haftung für evtl. verbliebene fehlerhafte Angaben und deren Folgen.

Alle Rechte vorbehalten

© Anchor Academic Publishing, Imprint der Diplomica Verlag GmbH
Hermannstal 119k, 22119 Hamburg
http://www.diplomica-verlag.de, Hamburg 2016
Printed in Germany

TABLE OF CONTENTS

Chapter 1 – Introduction 5
   1. 1. Arbuscular mycorrhiza 6
   1. 2. Classification and phylogenetics 7
   1. 3. Ecology of AM fungi 8
   1. 4. Benefits of AM fungi for the host plant 9
      1. 4. 1. Host plant growth 9
      1. 4. 2. Mineral nutrition 9
      1. 4. 3. Host biochemical composition 11
   1. 5. Screening of efficient AM fungi 13
   1. 6. AM association in oilseed crops 17
      1. 6. 1. AM association in sesame 17
   1. 7. Factors influencing AM colonization 21
      1. 7. 1. Edaphic factors 21
      1.7. 2. Host characteristics on AM symbiosis 25
   1. 8. Farming practices and AM symbiosis 27
   1. 9. Interaction of AM fungi with free-living nitrogen fixers 31
      1. 9. 1. Interaction with *Azospirillum* 32
   1. 10. Carrier based AM inocula and their field application 34
   1. 11. Aim and scope of the present study 35

Chapter 2 – Materials and Methods 37
   2. 1. Survey, collection, identification and maintenance of AM fungi from sesame soils 37
      2. 1. 1. Field site 37
      2. 1. 2. Sampling procedure 37
   2. 2. Studies on the development of fungal structures in sesame inoculated with indigenous AM fungi 39
   2. 3. Screening of efficient AM fungi for sesame 39
   2. 4. Investigation on the infectivity of AM fungus *G. dimorphicum* in sesame after long-term incubation in different soil types under two P levels 40

2. 5. Screening of sesame germplasm accessions for AM colonization and its influence on growth and P nutrition 41

2. 6. Evaluation of the effect of AM fungi, and P fertilization on mycorrhizal colonization growth, biochemical composition and nutrition of sesame under two water regimes 41

    2. 6. 1. Experimental 41

    2. 6. 2. Land preparation and evaluation of soil properties......................42

    2. 6. 3. Mycorrhizal inoculation...42

    2. 6. 4. Sowing and cultural practices 42

    2. 6. 5. Sampling 43

2.7. Studies on the interaction of AM fungi and *Azospirillum* in the rhizosphere of sesame under varied fertilizer level...............................43

2. 8. Identification of an efficient inoculation method for field grown sesame 44

    2. 8. 1. Study site and soil characteristics 44

    2. 8. 2. Crop and culture practices 44

    2. 8. 3. Experimental treatments 44

    2. 8. 4. Sampling 46

2. 9. Analytical methods 47

    2. 9. 1. Mycorrhizal colonization 47

    2. 9. 2. Soil analyses...49

    2. 9. 3. Plant analyses...49

2. 10. Statistical analysis 54

Chapter 3 – Arbuscular Mycorrhizal Associations with Field Grown Sesame 56

    3. 1. Characteristics of sesame soils..56

    3. 2. AM colonization 56

    3. 3. Spore density in soil...60

    3. 4. AM fungal diversity...62

    3. 5. AM fungi and nutrient range..68

    3. 6. Conclusions..69

Chapter 4 – Development of Fungal Structures by Indigenous AM Fungi during the Growth of Sesame 70

    4. 1. Conclusion 74

Chapter 5 – Relative Efficiency of Different Indigenous AM Fungi on the Growth and Yield of Sesame 75

    5. 1. Morphological characteristics 75

    5. 2. Biomass production 79

    5. 3. Yield components 81

    5. 4. AM colonization 84

    5. 5. Conclusion 86

Chapter 6 – Infectivity of Spores of AM Fungus *Glomus dimorphicum* on Sesame Plants after Long Term Incubation in Different Soil Types under Two P Regimes 87

    6. 1. Conclusion 90

Chapter 7 – Variability in natural infection by AM fungi and its influence on the growth and P nutrition of sesame accessions 91

    7. 1. Conclusion 94

Chapter 8 – Effect of AM inoculation, P fertilization and Irrigation on Growth, Yield, Biochemical Composition and Nutrition of Sesame 95

    8. 1. AM colonization 95

    8. 2. Growth and biomass production 97

    8. 3. Yield and quality attributes 103

    8. 4. Biometric parameters 110

    8. 5. Biochemical characteristics 116

    8. 6. Tissue P content 121

    8. 7. Conclusion 123

Chapter 9 – Individual and Interactive Effects of AM fungus *G. dimorphicum*, *Azospirillum* and Fertilizer in the Rhizosphere of Sesame and its Effect on Growth, Yield and Nutrition 124

    9. 1. AM colonisation 124

    9. 2. Growth characteristics 126

    9. 3. Yield and quality attributes 127

    9. 4. Plant nutrient status 130

    9. 5. Conclusion 132

Chapter 10 – Efficacy of Different AM Inoculation Methods on Field Grown Sesame 133

    10. 1. AM colonisation 133

    10. 2. Growth characteristics 137

    10. 3. Yield attributes 142

    10. 4. Conclusion 147

Chapter 11 – General Discussion and Conclusion 148

References 152

About the Authors 195

CHAPTER 1

Introduction

Sesame (*Sesamum indicum* L., Fam. Pedaliaceae) is recognized as the most ancient oil seed according to some archeological findings (Nayar, 1984; Bedigian and Harlan 1986). Its cultivation goes back to 2130 BC (Weiss, 1983). Sesame is cultivated in tropical, subtropical and southern temperate regions of the world for its seeds which is a rich source of edible oil. Sesame by virtue of its excellent quality oil is called the queen of the oil seed crops. Sesame seeds have the highest oil content (35-63%) among oilseed crops (Ashri, 1998; Baydar et al., 1999). The oil is very stable due to the presence of a number of antioxidants such as sesamin, sesamolin and sesamol (Suja et al., 2004). Therefore, it has a long shelf-life and can be blended with less stable vegetable oils to improve their stability and longevity (Chung et al., 2004; Suja et al., 2004). Recent studies have shown that the oil lowers cholesterol levels and hypertension in humans (Lemcke-Norojarvi et al., 2001; Sankar et al., 2004) and reduces the incidence of certain cancers (Hibasami et al., 2000; Miyahara et al., 2001). The observed effects have been attributed to the chemical composition of the oil characterized by a low level of saturated fatty acids and presence of antioxidants. The grains of sesame are eaten as fried, mixed with sugar or jaggery in the form sweets meats. Oil cake of sesame is a rich source of protein, carbohydrate and mineral nutrients such as calcium and phosphorus and is eaten avidly by humans as well as cattle.

India ranks both in the area and production of sesame in the world with an annual area of 2.07 million hectares and total production of 0.76 million tonnes (Anonymous, 2009a). In the state of Kerala, the crop is cultivated under an area of 533 hectares with an annual production of 171 tonnes of seeds (Anonymous, 2009b). The crop is mainly confined to the districts of Alappuzha, Palakkad and Kollam where it is cultivated in both uplands and lowlands. In the Onattukara region of Kerala, it is mainly cultivated as a summer crop in low land rice fallows.

Even though, sesame is the predominant oil seed crop of India, the per hectare productivity and the economic returns given by it are very low. It is due to the fact that the crop is very sensitive to biotic and abiotic stresses and it grows in marginal light-textured inceptisols having poor soil fertility associated with imbalance and or without fertilizer application. Under such unpredicted situations, the practice of application of both organic and inorganic fertilizer could help in bringing in profitable returns. However, due to escalating cost of production of chemical fertilizers and low subsidies for farmer, the agricultural planners are compelled to reorient their thinking towards cost effective and cheap renewable resources to supplement the chemical fertilizers.

Biofertilizers such as arbuscular mycorrhizal (AM) fungi have a potential to improve the sustainability of commercial sesame production by improving growth and yield. Such improvements of crops at given levels of inputs increases production efficiency and consequently reduce input levels to achieve the same yield. Reducing input levels can help in addressing some of the core issues of sustainability, such as eutrophication of waterways caused by excessive application of soluble P fertilizers and the depletion of non-renewable resources like rock phosphate. Improving the quality of crops by increasing their nutrient status also improves the sustainability of commercial agriculture in a less tangible, but equally important way, since the main goal of agriculture is to provide for the well being of human populations.

1. 1. Arbuscular mycorrhiza

Arbuscular mycorrhiza are mutualistic symbioses between plant roots and fungi belonging to the phylum Glomeromycota. The plant provides carbon to the fungal partner while the fungus improves the water and nutrient uptake from the soil.

Arbuscular mycorrhizal (AM) fungi have long been considered obligate symbionts with plants, since growing of the AM fungus without a host plant has not been possible. AM spores can germinate and produce hyphae, but they, die in the availability of a host. However, this view has been challenged by an experimental study showing that AM fungi can grow and form spores *in vitro*, if supplemented with a carbon source and stimulated by particular bacterial strains (Hldebrandt et al., 2006).

Spores are asexual, multinucleate structures that are produced directly by the mycelium, either inside or outside the root. In some species sporocarps are produced, where several spores get surrounded by a periderm like structure.

The hyphae are aseptate and can grow both outside (extraradical) and inside the roots (intraradical). The intraradical mycelium produces highly ramified structures called arbuscules inside the cortical cells of roots (*Arum*-mycorrhizal type). In some other cases, hyphal coils are formed instead (*Paris*-mycorrhizal types)

Many species of Glomeromycota also produce large intraradical, globose storage cells intracellularly called vesicles. Because of this glomeromycotan fungi are sometimes also referred to as vesicular-arbuscular mycorrhizal (VAM) fungi.

1. 2. Classification and phylogenetics

Dangeard (1900) was the first to name a VAM fungus. He isolated and described a typical VA mycorrhizal from poplar and named it as *Rhizophagus populinus*. Peyronel (1924) was the first to recognize the VAM fungi as *Endogone* species. Thaxter (1922) was the first to write a monograph of the family Endogonaceae where he described seven genera in their monograph of Endogonaceae viz., *Endogone*, *Glomus*, *Gigaspora*, *Modicella*, *Graziella*, *Acaulospora* and *Sclerocystis*. From time to time, new members were added to the family Endogonaceae by different workers. Schenck and Perez (1987) reported 120 species of soil fungi forming VA mycorrhizal under the genera *Glomus*, *Gigaspora*, *Acaulospora*, *Scutellospora*, *Entrophospora* and *Sclerocystis*. These genera were distinguished from each other on the basis of the manner of spore formation by them. The new order Glomales which included 'all soil borne fungi which form arbuscules in obligate mutualistic associations with terrestrial plants' separated AM fungi from others in the Endogonales an order in which included a group of saprophytes which may also produce ectomycorrhizae like associations. The order Glomales thus included three families, the Acaulosporaceae, Gigasporaceae and Glomaceae with six genera: *Acaulospora*, *Entrophospora*, *Gigaspora*, *Glomus*, *Sclerocystis* and *Scutellospora* (Morton and Benny, 1990).

Traditional taxonomy of AM fungi has mainly been based on spore morphology and ontogeny. The structures and characters of the mycelia, e.g. vesicles and arbuscules and coils are of great taxonomic importance. The application of molecular techniques to identify AM fungi in field studies has uncovered large cryptic diversity of new AM fungi that are not taxonomically characterized. Complete phylogenies in the Glomeromycota are based on the 18S rRNA gene (Schüßler et al., 2001). Morton and Redecker (2002) reported two ancestral clads of species discovered from deeply divergent ribosomal DNA sequences. They are classified into two new families Archaeosporaceae and Paraglomaceae. At the present time, each family consists of one genus *Archaeospora* including three species forming a typical *Acaulospora* like spores from soporiferous saccule and *Paraglomus* consists of two species forming spores indistinguishable from those of *Glomus* species.

1. 3. Ecology of AM fungi

AM fungi are of worldwide distribution. They are present in the soil in the form of chlamydospores, zygospores and azygospores and have been recovered from the soil of a variety of habitats, e.g., nutrient deficient soils (Hayman et al., 1976), forests (Mohankumar and Mahadevan, 1988), deserts (Malibari et al., 1988), sand dunes (Gemma et al., 1989), savanna (Dodd et al., 1990), saline marshy areas (Sengupta and Chauduri, 1990), eroded soils (López-Sánchez and Honrubia, 1992) and contaminated soils (Bindu and Harikumar, 2008).

AM fungal association occur widely throughout the plant kingdom (Gerdemann, 1968). They have been reported to be present in bryophytes, pteridophytes, gymnosperms and angiosperms in nearly all the geographical regions of the world (Agarwal, 2005). Most of the field crops belonging to diverse families have been shown to form mycorrhizal associations (Barea, 2000). Likewise, many of the plantation crops (Karunasinghe et al., 2009) and medicinal plants (Gupta et al., 2009) have been shown to form mycorrhizal associations.

1. 4. Benefits of AM fungi for the host plant
1. 4. 1. Host plant growth

Progress in the study of the function of the symbiosis was made in Europe in 1957, when Mosse published a report showing that arbuscular mycorrhizal infection led to improved growth of apple seedlings and clonal leaf bud cuttings. She used sporocarps of *Endogone* (*Glomus*) *mosseae* to inoculate plants growing in autoclaved soil. In 1958, Peuss showed that inoculation with mycorrhizal roots increased growth of tobacco growing in subsoil or in soil that had been fallow. In 1963 Clark, using surface-sterilized mycorrhizal roots as inoculum, reported an increase of growth of tulip poplar trees planted in fumigated soil, and Meloh (1961, 1963) showed that the growth of maize and oats could be improved by AM fungi. Various laboratory and greenhouse experiments have demonstrated that AM inoculation can improve growth and nutrition of crop plants. Khaliel and Elkhider (1987) observed that the tomato plants inoculated with *Glomus mosseae* showed greater dry weight and higher percentage of survival than uninoculated plants in low P soil. In a study of the beneficial effect of AM fungi on tuber crops, Potty (1993) and Potty and Harikumar (1995) observed an increased growth and tuber yield in sweet potato due to inoculation with the AM fungi *Glomus microcarpum*. Mohammad et al. (1998) observed that AM colonization resulted in an increased growth, dry matter production and grain yield in field grown wheat. Enhanced shoot and fruit dry weight and total dry biomass in mycorrhizal chile anchopepper has been reported (Auguilera – Gómez et al., 1999). A similar increase in growth characters due to mycorrhizal association has been reported in many crops such as subterranean clover (Grazey et al., 2004) strawberry (Stewart et al., 2005) and green gram (Idnani and Singh, 2008).

1. 4. 2. Mineral nutrition

Most research on AM effects on plant nutrition has been concerned with phosphate because it is one of the major plant nutrients. It is now well established that mycorrhizae can improve the P nutrition of host particularly in low to moderate fertility due to the exploration of the soil by the external hyphae beyond the root hairs and P depletion zone. Absorbed P is probably converted into polyphosphate granules in the external hyphae (Callow et al., 1978) and passed to the host (White and Brown, 1979). This flow of P occurs in the presence of acid phosphatases (Gianinazzi et al., 1979) during the arbuscule life span (Cox and Tinker, 1976) or senescence (Kinden and Brown, 1975).

Experiments conducted in unsterilized soils have frequently shown that introduced fungi could stimulate more P uptake than indigenous AM fungi (Islam et al., 1980). Similarly, increased plant growth responses to AM inoculation in soils incorporated with rock phosphate and decreased plant growth in soil applied with super phosphate have been observed (Jalali and Thareja, 1985; Tang and Chen, 1986). Increased P uptake due to AM inoculation has been reported by several workers (Diop et al., 2003; Caglar and Bayram, 2006).

Higher concentrations of nitrogen in tissue have been reported in AM associated plants (Hawkins et al., 2000; Azcon et al., 2001). Habte and Aziz (1985) observed increased nitrogen uptake in mycorrhizal *Sesbania grandiflora*. Smith and Read (1997) reported that the increase in total nitrogen is due to the higher P uptake through the AM hyphae rather than increased soil uptake. Oliver et al. (1983) found that higher uptake of phosphorus in mycorrhizal plants increases the activity of NAD dependent nitrate reductase enzyme both in shoot and root, hence, the possibility of mycorrhizal fungi having a NAD dependent enzyme which might contribute to nitrate reductase activity.

Mycorrhizal effects on N nutrition have been studied under field conditions and the potential of increased uptake of N from soil, as well as P mediated effects on N fixation have been demonstrated in *Hedysarvum coronarium* (Barea et al., 1987). In mixed plantings, a twofold increase in ^{15}N transfer from soybean to maize has been observed in mycorrhizal plots together with the relative increase in productivity of maize (Hamel et al., 1991). This suggested that mycorrhizal fungi may be involved in the redistribution of N in the plant community.

In many investigations, K was found to be at higher concentrations in the tissues of mycorrhizal plants than in those of non-mycorrhizal plants (Nielsen and Jensen, 1983; Diop et al., 2003). Elevated concentrations of K in shoots, not in the roots of mycorrhizal *Trifolium subterraneum* plants grown in P deficient soils were observed (Smith and Walker, 1981). Huang et al. (1985) reported higher K uptake in mycorrhizal Leucaena *leucocephala* and K depletion in a hyphal compartment colonized by *G. mosseae* and increased accumulation in associated mycorrhizal *Agropyron repens* has been observed (George et al., 1992).

The uptake of micronutrients such as Zn, Ca, Mg and Fe were also found to be enhanced by AM inoculation (Bagyaraj and Manjunath, 1980; Cooper and Tinker 1981; Lu and Miller 1989; An et al., 1993; Marschner and Dell, 1994; Clark et al., 1999).

1. 4. 3. Host biochemical composition

AM fungi have fundamental effects on host plant biochemistry and physiology (Smith and Gianinazzi-Pearson, 1988; Koide and Schreiner, 1992). The increased concentration of soluble carbohydrates in host root tissue attributable to mycorrhizal colonization has been observed (Steffens et al., 1963; Thomson et al., 1986; Khalafallah and Abo-Ghalia, 2008). Significant amounts of carbohydrate trehalose was found in the root of some host fungus combinations (Schubert et al., 1992) which contradicts the earlier findings of Hayman (1973) that there is no indication of fungal carbohydrate such as trehalose and mannitol. Total CHO pools increased with mycorrrhizal dependency of citrus genotypes providing evidence that C allocation pattern of the host affects mycorrhizal colonization (Graham et al., 1997).

Increased P supply decreased the per cent root length converted to mycorrhizas and concentration of soluble carbohydrate (Same et al., 1983). The hypothesis that high concentration of P inhibits the formation of VA mycorrhizas by reducing concentration of soluble carbohydrate in the root was tested on leek by Amijee et al. (1990) found that even at the concentration of soil P at which the infection was reduced, the concentration of soluble carbohydrate increased to its maximum discounting the above hypothesis.

Total amino acid levels in roots was found to be lower in mycorrhizal leek (*Allium porrum*) plants (Rolin et al., 2001) however, an optimal amino acid level, especially of glutamate necessary for growth of AM fungi within the root cells (Tawaraya et al., 1990). In a study of the efficacy of AM fungus *Glomus fasciculatum* towards amino acid levels in *Prosopis cineraria* under glass house condition, a total of twelve amino acids were quantified in mycorrhizal and non mycorrhizal plants. AM inoculation resulted in increased levels of all the amino acids as compared with control (Mathur and Vyas, 1996). Higher values of free amino acids were noted in tea leaves when grown in pots inoculated with *Glomus fasciculatum* and *Gigaspora margarita* (Lin et al., 2006). Recently, Matsubara et al. (2008) determined that inoculation of AM fungi increased total amino acid concentration and concentration of specific amino acids in strawberry plants.

The accumulation of α amino acids like proline in plant tissue is influenced by various abiological stresses. Drought stress is known to result in decline of AM colonization, but the proline level of AM plants was higher than non-mycorrhizal controls during drought period (Goicoechea et al., 1998; Valentine et al., 2006). The similar effect of AM fungi on stomatal conductance, photosynthesis and proline accumulation was observed in *Citrus jambhiri* by Levy and Krikun (2006). Further, Lioussanne et al. (2008) noted a significantly higher concentration of proline and isocytrate in the roots of tomato inoculated with *Glomus intraradices*.

Differences in qualitative and quantitative expression of protein have been shown in AM fungi (Dumas et al., 1989; Wyss et al., 1990; Arines et al., 1993; Arines et al., 1994). It seems that mycorrhiza formation increases the expression of low molecular weight protein as suggested by the results obtained in soybean (Pacovsky, 1989) tobacco (Dumas et al., 1990) and mulberry (Kumaresan, 1997).

Protein content was much higher in mycorrhizal than in non-mycorrhizal root extracts, in tobacco and onion (Dumas et al., 1989). A two to six fold higher protein content was found in mycorrhizal than in non-mycorrhizal red clover roots (Arines et al., 1993). Other reports have not shown such a large difference in both types of roots as have been described so far, and further deep studies of this aspect are needed. Perhaps this difference is a consequence of facts such as higher metabolic activity in AM colonized root cells and the presence of internal and external fungal mycelium. It is difficult to speculate further because our actual knowledge of AM fungal proteins is very limited, and we do not know if the new proteins are of fungal (or) plant origin.

1. 5. Screening of efficient AM fungi

Use of AM fungi for the improvement of crop productivity requires selection of an efficient and appropriate fungus (Menge, 1983) since AM fungi differ in their ability to from efficient symbiosis with different crop plants. AM fungi found to be efficient for some field crops are depicted in Table 1.1.

Table 1.1. List of efficient AM fungi for some field crops

Plant species	Efficient AM fungus/ fungi	AM fungi tested	Parameters evaluated	References
Allium cepa	*Gigaspora margarita*	*G. margarita, G. calospora* and a known AM fungus	Leaf number dry matter P uptake, bulb yield	Ramana and Babu, 1999
Amorphophallus paeonifolius	*Glomus mosseae* + *G. aggregatum*	*G. mosseae, G. aggregatum, Gigaspora albida, Pisolithus tinctorius*	Tuber yield /plant	Ganesan and Mahadevan, 1994
Arachis hypogeae	*Glomus* sp.	*G. etunicatum, Gigaspora margarita*	Shoot dry matter, nutrient uptake	Ahiabor and Hirata, 1994
	Glomus fasciculatum	*G. fasciculatum* and seven other AM fungi	Root shoot length dry weight P content, root colonization	Vijayakumar and Bhiravamurthy, 1999
Cajanus cajan	*Glomus clarum*	*G. fasciculatum* and seven other AM fungi	Plant growth	Diederichs, 1992
Capsicum	*Glomus intraradices*	*G. intraradices*, indigeneous mixed culture, commercial inoculum (Mycorise & C)	Fruit yield	Gaur et al., 1998
Colocasia esculenta	*Glomus mosseae* + *G. aggregatum*	*G. mosseae, G. aggregatum, Gigasproa albida, Pisolithus tinctorius*	Tuber yield/plant	Ganesan and Mahadevan, 1994

14

Cucumis sativus	Glomus caledonium	G. caledonium, Glomus sp.	P uptake	Joner and Jakobsen, 1994
Eleusine coracana	Glomus caledonium	G. caledonium, G. mosseae, G. fasciculatum, G. epigaeum (G.versiformae), Gigaspora calospora, G. margarita	Mycorrhizal efficiency root colonization	Tiwari et al., 1993
Lycopersicum esculentum	Glomus etunicatum	G. etunicatum, G. mosseae	Shoot dry weight, plant height	McGraw and Schenck, 1981
Manihot esculenta (cassava)	Glomus fasciculatum	Glomus fasciculatum, G. mosseae G. constrictum, G. etunicatum, Acaulospora morrowea	Root colonization plant weight, shoot and root dry weight	Sivaprasad et al., 1990
	G. mosseae + G. aggregatum	G. mosseae, G. aggregatum Gigaspora albida, Pisolithus tinctorius	Tuber yield / plant	Ganesan and Mahadevan, 1994
Oryza sativa cv Prakash	Glomus intraradices	G. intraradices, G. fasciculatum	Grain yield	Secilia and Bagyaraj, 1994

Crop	Inoculum	Fungi tested	Parameters	Reference
Oryza sativa (upland rice)	Acaulospora spinosa	Acaulospora spinosa, A. scrobiculata	Plant biomass grain yield, root colonization	Ammani and Rao, 1996
Phaseolus mungo (mung bean) (Vigna radiata)	Glomus intraradices	Glomus intraradices, Acaulospora scrobiculata and four other fungi	Plant growth, crop yield	Vasuvat et al., 1987
Triticum aestivum wheat (var.swift)	Glomus intraradices	Glomus intraradices, Gigaspora margarita	Plant yield, number of grains/spike	Asif et al., 1995
Vigna unguiculata	Glomus etunicatum	G. etunicatum, Gigaspora margarita	Shoot dry matter nutrient uptake	Ahiabor and Hirata, 1994
Zea mays (pioneer 3905)	Glomus etunicatum	G. etunicatum, G. mosseae G. aggregatum, G. versiformae	Leaf mass, protein concentrations	Boucher et al., 1998
Vitis vinifera	Acaulospora laevis and a mixed inocula	Acaulospora laevis, A. scrobiculata, Entrophospora colombiana, Gigaspora gigantea Glomus manihotis, Scutellospora heterogama and a mixed AM inocula	Survival and growth	Harekrishna et al., 2006

1. 6. AM association in oilseed crops

Arbuscular mycorrhizal fungi are known to occur abundantly in tropical soils supporting oilseed crops (Manoharachary and Prakash, 1991). Groundnut is perhaps the most studied oil seed crops with reference to mycorrhizal symbiosis. Investigations on growth and nutrient uptake response of this crop to AM inoculation have indicated an increase in growth and uptake of nutrients such as P, Zn and Cu due to mycorrhizal inoculation in comparison to uninoculated controls (Daft and Elgiahmi, 1976; Rao and Parvathi, 1982). Concurrent changes in the biochemical constituents such as carbohydrates and proteins in groundnut have been noticed due to AM inoculation (Krishna and Bagyaraj, 1984). A preliminary survey made by Sulochana and Manoharachary (1989) has revealed the association of *Glomus constrictum*, G. *fasciculatum and G. mosseae* with the rhizosphere soil and root region of safflower under field conditions grown in nutrient deficient soils with less moisture.

Perennial plants that produce oil seeds such as coconut and oil palm are not an exception for AM association (Girija and Nair, 1985; Thomas and Ghai, 1987; Blal et al, 1990) and the beneficial effect of AM association on the growth and nutrition of these crops has also been reported (Harikumar and Thomas, 1991; Widiastuti and Tahardi, 1993).

1. 6. 1. AM association in sesame

Incidence of AM colonization and its beneficial effect on the crop has been examined by several workers. An overview of AM studies carried out in sesame is depicted in Table 1. 2.

Table 1. 2. An overview of AM studies carried out in sesame

Parameter	Growing situation	Results achieved	Reference
AM colonization	Field experiment	Extensive root colonization by AM fungi, which increased up to 6 weeks. Number of vesicles increased with age of the plant. Incidence of *Glomus*, *Gigaspora* and *Scutellospora* species in the rhizosphere	Vijayalekshmi and Rao, 1988
	Field experiment	Occurrence of AM fungi in the roots and rhizosphere of sesame. *Glomus macrocarpum*, *G. fasciculatum* and *Sclerocystis sinuosa* were the most prevalent species. The appressoria, arbuscules and vesicles in the roots of sesame were examined under SEM.	Selvaraj and Subramanian, 1988
	Field experiment	Fungicides carbendazim and blitox when applied to soil after 30 days of seed sowing significantly inhibited mycorrhizal colonization	Vijayalakshmi and Rao, 1993

	Field experiment	Genotype dependent variation in AM colonization of the crop. Predominance of *Glomus* spp.	Pavankumar et al., 1998
Spore density in soil	Field experiment	AM propagule number was more in the rhizosphere soils of *Kharif* crop than *rabi*. AM fungi belonging to *Acaulospora*, *Glomus* and *Gigaspora* were found associated with both seasons	Sulochana et al., 2000
Biochemical	Pot study	Inoculation of AM fungi resulted in a significant enhancement of phenolic content of plants in sterilized soil. Accumulation of lipid and phenolic compounds in AM structures; particularly neutral lipids and catechol tannins in vesicles	Selvaraj and Subramanian, 1990
	Field experiment	Inoculation of AM fungi at 45 kgh^{-1} rock phosphate showed an increased biomass and yield	Prakash and Tandon, 2002

Growth and yield	Pot study	Mycorrhizal plants showed a significant increase in parameters such as shoot and root dry weight, number of capsules plant^{-1} and capsule dry weight	Prakash et al., 2004
	Pot study	Inoculation of AM fungi increased dry biomass of plants	Leye, 2006
	Pot study	Inoculation with AM fungi significantly increased leaf number, leaf area and dry weight of roots	Boureima et al., 2007
	Field experiment	AM inoculation led to an increased P uptake by the crop	Prakash and Tandon, 2002
Nutrient status	Pot study	Mycorrhizal plant had a comparatively high N and P content in tissue	Prakash et al., 2004

1. 7. Factors influencing AM colonization

Distribution of AM fungi in soils and its development in plant roots is influenced by both edaphic and host characteristics.

1. 7. 1. Edaphic factors
1. 7. 1. 1. Soil type

Population of AM fungi varies in different soil types. Kruckelman (1973) noted that the loam soils contained significantly more spores than the sands. Similarly, the silty clay also contained more spore numbers than sandy soil. Kehri et al. (1987) reported highest population of *Glomus* in the sandy clay loam, while the lowest was in clay loam. In a survey of the occurrence of AM fungi in soils of Tamil Nadu, India Srinivas et al. (1988) observed highest spore count in sandy loam soil followed by laterite clayey and sandy. However, the highest percentage colonization in plant roots was noticed in plants grown in clayey soils followed by sandy loam and laterite. Red soil showed a spurt in both spore load and percentage root colonization in plants. Potty (1990b) reported a range of 80 – 90% root colonization in sweet potato grown in alluvial soil.

1. 7. 1. 2. Soil moisture

Soil moisture influences the extent of AM infection in the roots and distribution of spores in the rhizosphere (Mejstrick, 1972; Khan, 1974). Reid and Bowen (1979) reported that excessively high soil water potential reduces the infection by AM fungi. High moisture level in the field reduces oxygen tension of soils, reducing the development of the endophyte (Hayman, 1983). Mohankumar and Mahadevan (1986) reported that water logging may substantially reduce the number of spores in mangrove soils and may abolish mycorrhizal infection. Indications were given by Cerligione et al. (1988) and Al-Agely and Reeves (1995) on poor sporulation and root colonization in *Schizachyriun scoparium* and *Oryzopsis hymenoides* respectively due to high soil moisture. Nevertheless, rice cultivars remained colonized with AM fungi in their roots during their growth period under high moisture level and water logging condition (Gupta, 1996; Singh, 1996).

1. 7. 1. 3. Soil pH

The interaction between pH and colonization by different species of AM fungi appears to be complex. Graw (1979) found that colonization of Niger seed (*Guizotia abyssinica* [L.f.] Cass.) roots by *Glomus macrocarpum* was less at lower pHs (4.3 and 5.6) than at a moderate pH (6.6). Later, Abbot and Robson (1985) found that the effect of pH on AM colonization of roots varied with different species of AM fungi. In their study, AM colonization of subterranean clover roots by *Glomus fasciculatum* did not differ with soil pH modified by liming from 5.3 to 7.5 but colonization by *Glomus* sp. (WUM 16) was only extensive at the higher pH, and its hyphae were found not to grow from propagules in soil below pH 5.3. Similarly, Hayman and Tavares (1985) found that root colonization of strawberry (*Fragaria vesca* L.) by *G. fasciculatum* was unaffected by soil pH, but colonization of *G. mosseae* was greatly reduced in one soil with pH 4.0. However, the apparent sensitivity of some AM fungi to soil pH is not constant between all soils, nor is it limited to acidity. Hayman and Tavares (1985) found that the observed reduction in root colonization of strawberry by *G. mosseae* was limited to one of the two field soils they used and, furthermore, in the other soil, increasing soil pH reduced colonization of strawberry by *G. clarum*.

Differential sensitivity of AM fungi to soil pH may explain the findings of Wang et al. (1993), who utilized fields that were part of a long-term liming experiment. Wang et al. (1993) found that AM colonization of field grown oats and potatoes was unaffected by soil pHs ranging from 4.5 to 7.5. However, they recorded the presence of up to nine species of AM fungi, colonization by at least some of which could have been insensitive to pH like those used by Abbot and Robson (1985) and Hayman and Tavares (1985). Furthermore, at most of the soil pHs included in their experiments, the above workers found that root colonization by apparently pH-sensitive. AM fungi was reduced but not eliminated, which would not prevent them from proliferating if soil pH was favourably changed.

The effect of root colonization by various AM fungi at different soil pH levels on the dry weight (Rohyadi et al., 2004), P concentration, Zn concentration (Rohyadi et al., 2004) and yield (Wang et al., 1993) of various plants have also been reported in the above studies. However, given the wide variation in experimental conditions, useful comparison between these experiments is limited. Graw (1979) also found that the effect of pH on colonization differed with host species and that AM growth responses of plants appeared to be related to the form of pH supplied, which seemed to be differentially available to the plant or AM uptake pathway depending on substrate pH and the presence or absence of AM colonization.

Given the generally accepted view that the amount of colonization is not necessarily related to plant responses (Smith et al., 2003; Smith et al., 2004; Asghari et al., 2005) the effect of pH on AM colonization and subsequent growth responses is difficult to distinguish. It does seem clear that soil pH can influence the populations of many AM fungi and encourage or discourage species that may produce more or less positive growth response in different plants (Haas and Krikun, 1985; Wilson, 1988; Khaliel, 1993; McGonigle et al., 2003).

1. 7. 1. 4. Soil Ec

Arbuscular mycorrhizal fungi have been reported from saline soils (Rozema et al., 1986; Ho, 1987). These are associated with plant roots over a wide range of salinity levels and in varying habitats. Pond et al. (1984) surveyed AM fungi in 57 saline soil sites in 30 different plant species. In all these sites soil EC ranged from 1 dSm^{-1} to 199 dSm^{-1} and pH was alkaline. However, the number of AM fungal spores (Kim and Weber, 1985) and spore germination (Estaun, 1991; Juniper and Abbot, 1992) decreased as the concentration of Na in soil increased.

In saline soil growth of AM hyphae seems to be very slow due to the requirement of high energy for hyphal growth. This also depends upon maintenance of ionic balance in the mycelium and internal water potential and saline soils (Cook and Whipps, 1993). Colonization in roots after the primary infection is also influenced by NaCl (Gupta and Krishnamurthy, 1996; Mc Millan et al., 1998).

There are several reports on the role of AM fungi in salt stress conditions (Hirrel and Gerdemann, 1980; Dodd et al., 1990). Mycorrhizal inoculation of *Leucaena leucocephala* and *Prosopis juliflora* in saline soils diminished the adverse effect of NaCl or juvenile growth and enhanced seedling dry biomass, nutrient content and number of nodules (Dixon et al., 1993). Moreover, AM inoculation of *Sesbania aegyptiaca* and *Sesbania grandiflora* significantly increased growth of nursery raised seedlings in saline soil (Giri et al., 1999).

1. 7. 1. 5. Organic carbon

The studies of Warner and Mosse (1980) have indicated that AM fungi grow saprophytically in soils containing high level of organic carbon. Mohanakumar and Mahadevan (1988) however, did not find any relationship with organic carbon content, AM infection and spore count in a tropical reserve forest of Tamil Nadu. Hršelová et al. (1999) opined that AM fungi are associated with soil microsites rich in some easily mineralizable fraction of soil organic matter rather than with total or oxidizable organic carbon. The higher rate of organic carbon in the soil appears to promote taxon richness of AM fungi (Toljander, 2006).

1. 7. 1. 6. Nutrient status of soil

Several reports have shown that increasing concentrations of soluble phosphate in soils can decrease fungal colonization in roots. The reduction could result from direct inhibition of external hyphal growth (Sanders, 1975; Graham et al., 1981) or from indirect effects, including changes in the endomycorrhizal infection (Asimi et al, 1980; Plenchette et al., 1983; Schwab et al., 1983). However, where soil concentration of available phosphate is extremely low small additions of phosphate can positively influence infection.

There is much evidence that high concentration of soil P resulting in reductions in arbuscular mycorrhizal infections (Kahiluots et al., 2000) evaluated by trypan blue staining (Trouvelot et al., 1986; De Miranda et al., 1989).

The role of mycorrhiza in improving nitrogen fixation capacity has been well documented. But soil nitrogen has been reported to be negatively correlated to AM fungal spore number in soil (Hayman, 1975; Jenson and Jakobsen, 1980). Nitrogen in the soil both as NH_4^+ and NO_3^+ forms is known to reduce mycorrhiza development, NO_3 being more inhibitory than NH_4 (Menge, 1984; Alexander and Fairley, 1986).

Fertilization with K enhances spore production by AM fungi (Furlan and Bernier – Cardou, 1989), but AM plants often have depressed K concentrations (Siqueria and Paula, 1986). High nitrogen input in forest ecosystems seems to hinder mycorrhizal growth (Alexander and Fairley, 1983). Excess nitrate in the soil has a damaging effect on mycorrhizal growth (Robert, 1985).

1.7.2. Host characteristics on AM symbiosis
1.7.2.1. Host genotype

The extent of mycorrhizal infection in the roots and the response to inoculation with AM fungi vary with plant species (Menge et al., 1978). Different cultivars of a single plant species also differ in their ability to harbour AM fungi in their root system (Schenck et al., 1975). Mehraveran (1977) and Menge et al. (1978) reported genotype dependent variation in AM colonization of citrus. Mosse (1981) reported that genotype dependent variation in root colonization and response to AM inoculation would be due to the interaction between host genotype and AM strain preference. In a study of thirty genotypes of wheat across the field locations, Azcon and Ocampo (1981) observed a range of mycorrhizal colonization intensity between 25 and 56%. Similar genotype dependent variation in AM colonization was observed in Sorghum (de Fransca, 1981; Subhashini et al., 1988), barley (Tilak and Murthy, 1987), pea (Estaun et al., 1987) and capsicum (Suvercha and Mukerji, 1988) also. Krishna et al. (1985) reported that the genetic makeup and physiological need of a plant for P, determine the extent of colonization since, mycorrhizae play important role in P uptake. Genotype dependent preference for AM colonization was noted in plantation crops such as coconut (Thomas and Ghai, 1987). Genotype dependent variation was also observed in different cultivars of tuber crops such as *Manihot esculenta, Ipomoea batatas, Dioscorea alata, D. esculenta and Coleus* spp. intercropped with coconut (Sivaprasad et al., 1990). Graham et al. (1991) revealed that in high P soil, colonization varies substantially among plant families, genera and even among closely related genotype of the same species. Sureshkumar et al. (1995) on quantification of the mycorrhizal status of five varieties of pigeonpea (ICPL 84023, ICPL 85014, ICPL 88001, ICPL 88007 and ICPL 90012) observed a significant variation in percentage root colonization by AM fungi among the varieties. Harikumar and Potty (2002) screened 257 genetic stocks of field grown sweet potato for AM colonization and found that 20% of them responded to natural infection only to the tune of 25%, whereas another 20% had 50% infection in their root system and the remaining 60% had a very high level of colonization.

1. 7. 2. 2. Age of the host plant

The structure and function of mycorrhizal communities in response to plant age are a primary concern while utilizing mycorrhizal technology for the benefit of host plants (Lindermann and Hendix, 1982). Allen (1992) opined that the reduction of sporulation and colonization by AM fungi have in some cases, being related to plant age. Explanation of these variations includes the progressive death of the fungus within the root as plants age (Smith and Read, 1997). Some of the best evidence on the importance of host plant age on mycorrhizal communities has arisen from the studies of AM fungi conducted by Johnson et al. (2005) in tropical forests. Husband et al. (2006) also noted a decrease in AM fungal development as the age of the seedlings of *Tetragastris panamensis* increased.

1. 8. Farming practices and AM symbiosis

Many farming practices, including those used in conventional, commercial sesame production are well known to be deleterious to AM colonization or sporulation in a variety of crops (see Table 1.3). Perhaps the most obvious of these is soil cultivation, which appears to affect AM fungi in the soil in a number of ways. AM fungal inoculum in soils occurs naturally in two forms: (a) robust propagules (spores and colonized root fragments); and (b) fragile hyphae. One of the earliest studies on the effect of soil cultivation on AM fungi was conducted by Kruckelmann (1975), who used the frequency of spores in a given mass of dry soil to measure the effect of several tillage and fertilizer treatments on AM fungi. He found that tilling by rotary hoe had a negative impact on AM fungi compared with no tillage. Later work focused on AM colonization of plant roots as an appropriate measure and Jasper et al. (1989) concluded that soil cultivation reduced AM colonization of subterranean clover (*Trifolium subterraneum* L.) by damaging the AM hyphal network in the soil. Fairchild and Miller (1988) reached the same conclusion in their experiments on maize grown in pots subject to soil disturbance at regular intervals. This was confirmed by Evans and Miller (1990), who demonstrated that the soil disturbance – induced destruction of the AM mycelial network in root-free pot compartments were largely responsible for reduced AM colonization of maize growing in adjacent pot compartments. This appears to be supported by the field experiments of McGonigle and Miller (1999), who found that soil disturbance rather than reduced propagule production caused by low winter soil temperatures, reduced the ability of AM fungi to colonize early season maize. McGonigle et al. (1990) also speculated that disturbance to the external mycelium may be responsible for decreased plant growth in disturbed soils.

Table. 1. 3. Farming practices detrimental to AM fungi and AM colonization of crop plants grown in farmed soils.

Farming practice	How achieved	Crop plant/s	Measurement	Reference
Soil cultivation	Rotary hoe	Beet root and Oats	No. of spores	Kruckelmann, 1975
	Plough	Maize	AM colonization	Anderson et al., 1987
	Disturbing pot soil	Clover	AM colonization	Jasper et al., 1989
	Disturbing pot soil	Maize	AM colonization	Fairchild and Miller, 1990
	Plough	Maize	AM colonization	Mc Gonigle and Miller, 1993
	Plough	Maize	AM hyphal density and colonization	Kabir et al. 1997
	Plough	Maize	AM hyphal density	Kabir et al., 1998
	Sieving field soil	Maize	AM hyphal density and no. of spores	McGonigle and Miller, 1999
Fertilizer, application	KH_2PO_4	Maize	No. of spores	Daft and Nicolson, 1969a
	N, P, K, Mg and Na	Rye and Oats	No. of spores	Kruckelmann, 1975
	N, P & K	Rye, Wheat, Barley, Oats and Bean	AM colonization	Strzemska, 1975

	P	Maize	AM colonization	Anderson et al., 1987
	KH$_2$PO$_4$	Onion	AM colonization	Vosátka, 1995
	Organic (Oil cake)	Pigeon pea	AM colonization	Panja and Chaudhuri, 1999
	Manure and Mineral	Wheat, Vetch Ryegrass, Clover	AM colonization	Mäder et al., 2000

Regardless of whether it is the amount of intact external mycelium or density of AM fungal propagules that determines AM colonization following cultivation, AM colonization of crop roots may not be a reliable indicator of the total effect of soil tillage on AM fungal and any subsequent crop response. McGonigle et al. (1990) found that although the shoot dry mass and P concentration of maize increased with decreasing severity of soil disturbance, these were not accompanied by an increase in AM colonization. They concluded that the AM hyphal network in disturbed soil was responsible for the increased plant mass and P concentration rather than the amount of AM colonization in the roots. Douds et al. (1995) found that tillage treatment affected population groups of AM fungi differently. They found that spores belonging to *Glomus occultum* group were more prevalent in a maize/wheat (*Triticum aestivum* L) rotation without tillage, but spores belonging to the *Glomus* spp. and *G. etunicatum* groups were more prevalent when the soil was tilled. Jansa et al. (2002) also found that the tillage affected different AM fungal species differently. In differently tilled soils from a long term field experiment they found that non-*Glomus* AM fungi were less prevalent in field soils although *G. fasciculatum* was also found in untilled soils. However, in contrast to the work of Helgason et al. (1998) which found AM fungal diversity in tilled soils was severely reduced, almost to a single species (*G. mosseae*). Jansa et al. (2003) found that conventional tillage did not reduce AM fungal diversity in soils when compared to reduced tillage.

Chemical fertilizers may not always result in reduced AM colonization as was the case in the work of Bolan et al. (1984), where P application increased AM colonization in subterranean clover. However, this phenomenon was observed after applying relatively small amounts of P to a low P soil and such examples do not represent the majority of work in this area. Strzemska (1975) found that a higher rate of NPK fertilizer application reduced AM colonization of field-grown rye (*Secale cereale* L.), wheat (*Triticum aestivum* L.), barley (*Hordeum vulgare* L.), oats (*Avena sativa* L.) and faba bean (*Vicia faba* L.). Furthermore, Daft and Nicholson (1969a) found that the timing of fertilizer application might also effect AM colonization. In a pot experiment they applied small amounts of soluble P to maize over increasingly long periods with a resultant decrease in AM colonization and spore production. In a subsequent experiment, they also found that a single application of soluble P applied to maize earlier in the season depressed AM fungal spore production more than applications applied later in the season or not at all. However Vosátka (1995) found that the effect of soluble P on AM colonization of field-grown onion was dependant on AM fungal species and the presence and absence of irrigation, with no discernable pattern.

The effect of P on AM colonization can be explained by two mechanisms. Firstly, increased P has been shown to directly reduce the initiation of new AM colonizations and their subsequent growth within plant roots (Amijee et al., 1989a; Amijee et al., 1989b; Bruce et al., 1994). Bruce et al. (1994) found this direct effect of P on AM colonization to be particularly apparent in the early stages of AM colonization. Secondly, increased P increases plant root growth at a faster rate than the length of root colonized resulting in a decrease in the percentage of root length colonized (Smith, 1982; de Miranda et al., 1989; Bruce et al., 1994).

The cultivation status of the soil and the method of application of soluble P have also been shown to affect AM colonization. Jasper et al. (1979) found that AM colonization in unfertilized, uncultivated soils with low plant available P was more sensitive to P application than in adjacent fertilized, agricultural soils. In a related pot experiment they also found that banded P application to farmed field soil tended to suppress AM colonization more than mixed or top-dressed applications. Furthermore, both Strzemska (1975) and Mäder et al. (2000) found that although higher rates fertilizer generally reduced AM colonization, the effect was dependant on the host crop.

AM fungi respond markedly to organic, fertilization and substrate additives. In studies conducted at the Department of Agricultural Microbiology, University of Agricultural Sciences, Dharwad, India, on three genotypes of wheat (DWR-39, DWR-163 and DWR-187), Gaonker and Sreenivasa (1994) found that organic amendments to soil had a positive influence on the proliferation of *G. fasciculatum* and that farm yard manure (FYM) stimulated the greatest root colonization in artificially inoculated DWR 39 and DWR-187. Baby and Rao (1996) observed an increased percent colonization, infection intensity and arbuscule development in rice plants amended with green leaf manure.

Organic fertilizers may increase propagule density in the soil (Harinikumar and Bagyaraj, 1989; Baby and Rao, 1996). Field experiments with the application of composted animal manures in rotation crops of *Zea mays*/vegetable crops (*Spinacea oleraceae* and /or *Capsicum annum*)/small grain crops (*Avena sativa or Triticum aestivum*) showed that chicken litter/leaf compost and dairy cow manure/ leaf compost enhanced spore populations of two type groups of AM fungal species (*G. etunicatum* type and the general *Glomus* spp. group including *G. mosseae*) relative to those found in plots treated with raw dairy cow manure and conventional fertilizer (Douds et al., 1997). Muthukumar and Udaiyan (2000) found an increase in both roots AM colonization and spore number in field-grown cowpea (*Vigna unguiculata* (L) Walp) amended with FYM (cow dung, sheep manure) and green manure (sunhemp and pongamia).

1. 9. Interaction of AM fungi with free-living nitrogen fixers

Soil provides a favourable habitat for a large number of free-living, nitrogen fixing microorganisms, which do not enter into any organic union with plant roots, and thus, do not form any root nodules. AM fungi and these bacteria can interact synergistically to stimulate plant growth through a range of mechanism that includes improved nutrient acquisition and inhibition of fungal plant pathogen. These interactions may be of prime importance in sustainable low-input agricultural cropping system that relay on biological processes rather than agrochemicals to maintain soil fertility and plant health (Artursson et al., 2006)

1. 9. 1. Interaction with *Azospirillum*

Azospirillum exerts considerable influence on the efficiency of AM fungi and plant growth. Studies conducted at Departmento de Consejo Superior de Investigaceones Cientificas (CSIC), Granada, Spain, showed that roots from maize (*Zea mays*) and ryegrass (*Lolium perenne*) plants inoculated with *Azospirillum* reduced C_2H_2 but there was no significant effect of inoculation on nitrogen concentration in the roots. In non-mycorhizal plants, inoculation with *Azospirillum* resulted in greater dry matter production than was achieved by supplying nitrogen as a fertilizer, but this trend was reversed at the last harvest in maize. With mycorrhizal plants, *Azospirillum* stimulated the development of AM and was effective in improving the plant growth at the last harvest of rye grass. In mycorrhizal plants of maize, *Azospirillum* produced similar plant growth and nutrient uptake as compared to that achieved with nitrogen fertilizer. The dual inoculation of maize by *Azospirillum* and *Glomus* produced plants of similar size and nitrogen content and higher phosphorus content at the last harvest in comparison with those supplied with nitrogen and phosphorus (Barea et al., 1983).

Pot culture experiment conducted by Konde et al. (1988) on onion (*Allium cepa*) in P-deficient soil revealed that simultaneous inoculation with AM fungi (*Glomus* sp. or *Gigaspora* sp.) significantly enhanced the fresh and dry weights, and N and P uptake by shoots and bulbs over their corresponding inoculation with single cultures and controls. Similarly the dual inoculation of brinjal with *Azospirillum brasilense* and *Glomus fasciculatum* combined with the application of 75% of the recommended P fertilizer dosage gave the highest yield in field trial on a P-deficient soil (Indi et al., 1990).

Dual inoculation of Palmarosa (*Cymbopogon martinnii* var. *motia*) with *Glomus aggregatum* and *Azospirillum brasilense* increased the growth, yield, and oil content significantly over AM alone *Azospirillum* alone or uninoculated controls. *A. brasilense* also stimulated the AM colonization and increased AM spore population in the rhizosphere soil of palmarosa (Ratti and Janardhanan, 1996).

Interactions between AM fungi and *Azospirillum* in association with other microbial inoculants have also been tried by several workers. For e.g. coinoculations of the alfalfa (*Medicago sativa* L) plants with the associative – and/or the obligate nitrogen-fixing bacteria (*Azospirillum brasilense, Rhizobium meliloti*) and/or the AM fungus (*Glomus fasciculatum*) were evaluated in a pot experiment under controlled conditions by Biro et al. (2000) and found that the multilevel treatments with both the diazotrophs showed a synergistic effect for almost all of the tested parameters. Similarly AM fungi (*Glomus mosseae and Glomus deserticola*) as well as microbial - free inoculants used as phyto stimulators (*Azospirillum*) or as biological control agents of fungi (*Pseudomonas and Trichoderma*) have shown beneficial effects on the growth and health of maize plants (Vázquez et al., 2000) without altering mycorrhizal colonization (Russo et al., 2005).

In a study of the interactive effects of phosphate solubilizing bacteria (*Bacillus polymyxa*), N_2 fixing bacteria (*Azospirillum brasilense*) and AM fungus (*Glomus aggregatum*) on palmarosa, Ratti et al. (2001) found that all microbes inoculated together helped in the uptake of tricalcium phosphate, which is otherwise not used by the plants and their addition at 200 mg kg $^{-1}$ of the soil gave higher productivity to the plants.

Although these are many studies concerning interactions between AM fungi and bacteria, the underlying mechanisms behind these associations are in general not well understood and their functional properties still require further experimental confirmation.

1. 10. Carrier based AM inocula and their field application

For mass use of AM fungi an efficient delivery system to the site of their use is necessary. Soil based inocula involve a large amount of soil and high cost of transport to the field. A carrier system with light, inert inorganic/organic material is necessary for developing efficient delivery systems for AM fungal inocula. A number of materials such as expanded day (Dehne and Backhaus, 1986) and lignite slurry (Potty, 1990) have been tried as carriers.

Pellets containing mycorrhizal inoculum have been used successfully to inoculate plants with mycorrhizal fungi (Hall, 1979; 1980). Jung et al. (1981) reported a unique method of producing pelleted inoculum of AM fungi which is called polymer entrapped microorganisms.

The use of alginate gel as carrier for AM fungi has been described by Diem et al. (1981). They entrapped the culture of *Glomus mosseae* in an alginate gel and the inoculum thus prepared was designated as AEE (alginate entrapped endomycorrhizal) fungus. In a pot experiment using *Vigna unguiculata* both nodulation and growth of plants inoculated with AEE were similar to those of plants inoculated in the normal way with fresh infested roots. Vassilev et al. (2001) observed improved infective potential and colonization efficiency when *Yarrowia lipolytica* was co-entrapped with the AM fungi. Furthermore, intra radical forms of *Glomus* sp. (vesicles and mycelial fragments) entrapped in alginate gel when used as inoculum regenerated in alginate beads and the regenerated mycelium infected leek roots under controlled condition (Strullu and Plenchette, 1991).

Various methods are proposed for placement of AM inoculum in the fields. Placing inoculum in layers or pads beneath the seeds so that roots will penetrate the inoculum appears to be one of the most desirable methods of inoculating plants with AM fungi. Jackson et al. (1972) studied several different methods to inoculate corn and found that layering inoculum 5 cm under the seed was superior to placing inoculum around the seed or banding the inoculum alongside the seed.

Banding or side dressing with AM inoculum next to seedlings or seeds can be an effective way to apply mycorrhizal inoculum especially if the quantity of inoculum is limited. Using a tractor-drawn fertilizer bander, Ferguson (1981) successfully inoculated 3½ month old citrus by applying granular mycorrhizal inoculum in a band 15 cm deep and 15 cm from the seedling row.

Mixing mycorrhizal inoculum with soil may be the most natural method for inoculating plants, but it can be inefficient since it often requires large amounts of inoculum to obtain rapid infection. Broadcasting and rototilling mycorrhizal inoculum into citrus seedbeds gave maximum infection of citrus by mycorrhizal fungi as well as maximum growth responses (Ferguson, 1981).

Seed pelleting with mycorrhizal inoculum would be the easiest and most efficient method of inoculating plants with AM fungi if it would provide consistently good infection. In this method mycorrhizal sievings from a pot culture is mixed with 1% aqueous solution of methyl cellulose and poured over the seeds and dried (Ferguson, 1981). However, many authors (Mosse and Hayman, 1980; Ferguson, 1981) have expressed the opinion that mycorrhizal inoculum must be placed in the root zone to be effective. Seed pelleting with mycorrhizal fungi can be effective, but susceptible roots may pass through the zone of inoculum so rapidly that the infection may not take place.

For crops that will be transplanted pre-inoculation in the seedbeds is probably the most effective method for inoculation. Pre-inoculated plants are frequently used in mycorrhizal field experiments (Khan, 1972; Hayman and Mosse, 1979; Potty 1988; Harikumar and Potty, 2002).

1. 11. Aim and scope of the present study

Sesame crop is highly mycorrhizal and mycorrhiza dependent. In order to make use of the benefit of mycorrhizal technology to the maximum extent in crop production strategies, it is required to study the distribution pattern of the fungi in different ecosystems, its biology within the host and factors governing their incidence in soil and establishment in the host. Further, the study has to be extended to identify the efficient species/strains that enhance growth and nutrient uptake. Investigations on genotype dependent variation in AM colonization will lead to an understanding of the host-microbe interaction. A better understanding of the fluctuation of AM population and their performance under different cultural practices is essential for harnessing them for the benefit of the crop. The introduction of a third component into the mycorrhizal system-the N_2 fixing endophytes may further improve nutrient availability to enhance growth and yield of the crop. AM fungi being an obligate symbiont and difficult to grow it in axenic culture, the development of a mass multiplication and field inoculation strategy will be promising.

Keeping the foregoing account in view, the present investigation has been targeted with the following broad objectives.

- Survey, collection, identification and maintenance of AM fungi from different sesame growing areas.
- Studies on the development of fungal structures in sesame inoculated with indigenous AM fungi.
- Screening of efficient AM fungi for sesame.
- Investigations on the infectivity of the AM fungus in different soil types under sesame cultivation.
- Screening of selected germplasm accessions of sesame for natural AM colonization and its relationship with growth and P nutrition of the crop.
- Examine the effects of fertilizer and irrigation on AM colonization, growth and nutrition of the crop.
- Investigate the interaction of AM fungi and *Azospirillum* in the rhizosphere of sesame under varied fertilizer levels and its beneficial effects on the crop.
- Identification of an efficient inoculation method of AM fungi for field grown sesame.

CHAPTER 2

Materials and Methods

2. 1. Survey, collection, identification and maintenance of AM fungi from sesame soils

2. 1. 1. Field site

Studies were carried out in the major sesame growing regions of Kerala State (10^0 00'N, 25'E) South India, The sampling (S_1 to S_{18}) selected at random included regions of Kollam, Alappuzha and Malappuram districts of the State (Fig. 2.1). The climate is typical of tropical areas; mean annual air temperature is 28°C. Rainfall though erratically distributed sometimes, often tends to accumulate during June – July and September – October months of the year. Annual precipitation averages 247 mm (period 1999 – 2000). The soil types of the sampling sites are predominantly sandy and the soil characteristics are presented in Table 2.1.

Table 2. 1. Characteristics of the experimental soils

Soil type	Taxonomy	Soil pH	Organic C (g kg^{-1})	Soil P (mg g^{-1})
Coastal alluvium (CA)	Entisol	6.4	6.0	0.02
Greyish Onattukara (GO)	Entisol	6.3	5.4	0.03
Kuttanad alluvium (KA)	Inceptisol	3.2	7.7	0.01
Laterite (LA)	Oxisol	6.2	4.5	0.05

2. 1. 2. Sampling procedure

Sampling was done from the field with a standing crop at seven days between March 1999 and April 1999. Five plants at 10 – 15 m from each site were selected for sampling. From each test plant intact roots were collected. About 500 g of soil sample from the rhizosphere of each plant at a depth of 15 – 30 cm was collected in polythene bags and stored at 5°C till processing.

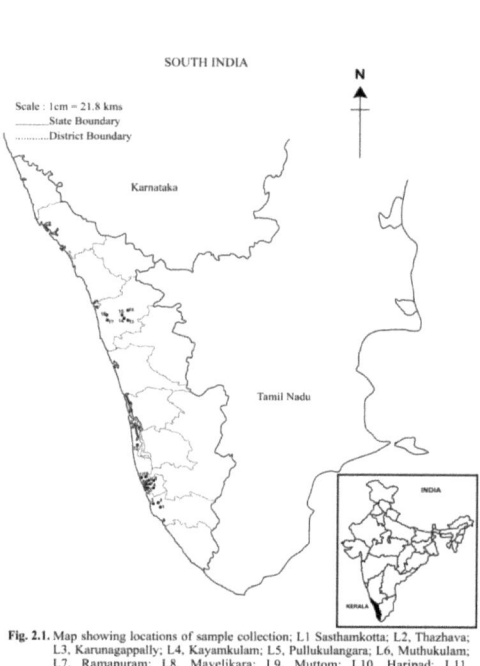

Fig. 2.1. Map showing locations of sample collection; L1 Sasthamkotta; L2, Thazhava; L3, Karunagappally; L4, Kayamkulam; L5, Pullukulangara; L6, Muthukulam; L7, Ramapuram; L8, Mavelikara; L9, Muttom; L10, Haripad; L11, Karthikappally; L12, Muhamma; L13, Melemury; L14, Pookothur; L15, Velliampuram; L16, Panthalloor; L17, Ponmunda; L18, Thalappara

2. 2. Studies on the development of fungal structures in sesame inoculated with indigenous AM fungi

Sesame plants (var. Tilatara) were grown in earthen pots (18 cm diameter) filled with sandy (Entisol) soil. The soil had a pH 5.6, Organic carbon 11.75 kg^{-1} and available P .75 mg g^{-1}. The soil was sterilized by autoclaving for 2 h prior to sowing the seeds. Five AM fungal isolates (*Acaulospora delicata, Acaulospora lacunosa, Glomus dimorphicum, Glomus versiformae, Scutellospora nigra*) procured from sesame soil multiplied and maintained on the sorghum plant for about 90 days were used in inoculation treatments. 50 ml of AM inoculum containing chlamydospores (approximately 200 spores) and infected root segments of *Sorghum* plant was placed 2 cm below the soil surface before sowing to produce mycorrhizal plants. After emergence of seedlings, the number of plants was thinned to one per pot and irrigated with equal quantities of deionized water as and when required. The experiment was set up in a glass house with five mycorrhizal treatments and one non-mycorrhizal control each replicated 21 times. Three plants were harvested from each treatment at 10 d intervals up to 70 d of growth and the roots were separated. The root samples were stored in 50% alcohol till further processing.

2. 3. Screening of efficient AM fungi for sesame

Sesame seeds (var. Tilatara) were sown in plastic pots (23 cm diameter) filled with 4 kg of autoclaved sandy (Entisol) soil (pH 5.6, organic carbon 11.7 g kg^{-1} soil and available P.7.5 mg g^{-1} soil). At potting AM fungal inocula of the isolates mentioned in item No.2.2 which consisted of roots and soil/sand mix of the culture were diluted to provide equal inoculum densities (approximating 1.7×10^5 propagules pot^{-1}) and placed uniformly 5 cm below the surface of the soil in all pots except control prior to sowing the seeds. Control plants received 100 ml of AM inoculum washing that had been passed through a Whatman 40 filter paper. After emergence of seedlings, the number of plants was thinned to one per pot and was given need-based irrigation with equal quantities of deionized water. The experiment was set up in a glass house with five mycorrhizal treatments and one non-mycorrhizal control each replicated nine times. The plants were harvested at 25 days interval up to 75 days of growth to measure various parameters. The sub-samples of root were stored in 50% alcohol till further processing.

2. 4. Investigation on the infectivity of AM fungus *G. dimorphicum* in sesame after long-term incubation in different soil types under two P levels

Four different soil types (Table 2.1) used for the study were collected within one week after harvesting the standing crops from fields where sesame cultivation was practiced for several years. Soils were air-dried and sieved through 2-mm mesh sieve. The soils were steam sterilized and filled in plastic pots of 10 cm dia having a holding of 1 kg soil. Native microflora (except AM fungi) was supplemented to the soil as sieving. Fifty ml of *G. dimorphicum* inoculum containing chlamydospores (approximately 200 spores) and infected root segments of *Sorghum* plant was uniformly mixed to the soil in all pots. Half of the pots received P fertilization as Rock phosphate (RP) @ 120 mg pot^{-1}. The mouth of the pots were covered with polythene sheets and tied properly to avoid the chance of contamination. The experiment consisted of four soil types (CA, GO, KA, and LA) and two levels of P (No added P (P_0) and added P (P_1) with a complete 4 × 2 factorial design with 18 replications per treatment. The 144 pots were incubated at 27^0 C in both light and dark in a glass house.

Three pots from each treatment were sown with sesame (var Tilatara) seeds at one-month interval. After the emergence of seedlings, the number of plants was thinned to one per pot and allowed to grow for a period of two months. The plants were irrigated with equal quantities of deionized water as and when required. The plants were harvested at two months growth to monitor root infectivity by AM fungi. The experiment was terminated at six months after soil incubation with AM fungi.

2. 5. Screening of sesame germplasm accessions for AM colonization and its influence on growth and P nutrition

Twenty accessions of sesame (courtesy KAU (RS) Kayamkulam) collected from different parts of the country were evaluated for AM colonization and its relationship with growth, yield and P content of the plants.

The study was conducted in a farmer's field located at Alappuzha, Kerala. The soil was sampled for analysis of physico-chemical properties and AM fungal population shortly prior to the initiation of the experiment. The soil was a sandy Entisol with a pH 5.5 (1:2:5; soil: water) and organic carbon 1.17%. Soil nutrient determinations included 108 kg Nh^{-1}, 25kg Ph^{-1} and 19.40 kg Kh^{-1}. The total indigenous AM fungal spore density prior to the start of the study was 206 spores per 100 ml soil consisting predominantly of *G. microcarpum, G.mosseae* and *G. margarita*.

Seeds were hand sown in rows at a distance of 1 ft with three replicate pits for each accession. After emergence of seedlings, the plant number in each pit was thinned to three to avoid overcrowding.

Plants were harvested at 75 days after emergence. Three replicated plants for each accession were dug out with almost the entire root system intact. The harvested plants were utilized for monitoring mycorrhizal colonization, growth and tissue P content.

2. 6. Evaluation of the effect of AM fungi, and P fertilization on mycorrhizal colonization growth, biochemical composition and nutrition of sesame under two water regimes

2. 6. 1. Experimental

A field experiment was conducted during 2000/2001 growing season on sandy (Entisol) soil to study the effect of AM inoculation and rockphosphate fertilization of growth and nutrition of sesame under two water regimes. The experiment was laid-down in an asymmetric ($2^2 \times 3 \times 5$) completely randomized design with two levels of AM fungi (uninoculated – M_0 and inoculated – M_1) two levels of water management (rainfed I_0 and irrigated I_1), three stages of growth (D_1-25 DAS, D_2-50 DAS and D_3 - 75 DAS) and five levels, of phosphorus (no added P – P0, 3.75 kgh^{-1} P-P1, 7.5 kgh^{-1} P-P2, 11.25 kgh^{-1} P-P3 and 15 kgh^{-1} P-P4). All the 60 treatments were carried out in triplicate.

2. 6. 2. Land preparation and evaluation of soil properties

A farmer's field located at Avoor, Kerala was chosen for conducting the experiment. The field was prepared by ploughing with a power tiller. The prior cropping system entailed rice production. Composite soil samples were taken to a depth of 30cm and analyzed for major soil properties and indigenous AM fungal spores. The soil properties before sowing was pH 5.2; organic carbon 1.43%; N 144 kgh^{-1}; P 20 kgh^{-1} and K 18 kgh^{-1}. Total indigenous AM fungal spore density in the soil was 170 spores per100 ml soil consisting of *A. delicata, G. mosseae* and *G. versiformae*.

2. 6. 3. Mycorrhizal inoculation

In mycorrhizal treatments, the inoculum of AM fungi *G. dimorphicum* was placed in the furrows below the seeds. The inoculum consisted of AM-colonized root pieces, spores and hyphae mixed with soil (2 kg per row), No inoculum was added to the non-mycorrhizal treatments.

2. 6. 4. Sowing and cultural practices

Seeds of sesame (cv. Tilatara) were sown by hand in rows directly over the inoculum in furrows and lightly covered with soil. There were five sesame rows in each block with spacing of 1 meter. When the plants reached to about 15 cm in height, the plant density was thinned to give a spacing of 15 - 25 cm between the plants. About 25 plants were maintained per row in a block. Nitrogen (urea) and potassium (muriate of potash) were broadcast to all plants irrespective of treatments and incorporated below the soil surface at a rate of 30 kgh^{-1}, Phosphorus (RP) was applied in treatments as graded level as mentioned earlier. In irrigated treatments, irrigation was given to the field capacity using a sprinkler at 15 days interval. Irrigation continued up to the time the pods begin to mature. Rainfed plants did not receive irrigation throughout the growth period. Weeds were controlled by hand as required.

2. 6. 5. Sampling

Three plants from each treatment were pulled out with the roots intact at 25 days interval for monitoring various parameters. Final sampling of plants from each treatment was done at the termination of the experiment at 75 days of growth.

2.7. Studies on the interaction of AM fungi and *Azospirillum* in the rhizosphere of sesame under varied fertilizer level

A pot culture study was conducted to evaluate the individual and interactive effects of AM fungi and *Azospirillum* in the rhizosphere of sesame under varied fertilizer levels. Experiment was carried out in earthen pots of 13 cm diameter having a holding of 5 kg soil. The pots were filled with sandy (Entisol) soil. The properties of the soil before the experiment were pH 5.5, organic carbon 11.78 g kg^{-1}, N 0.22, P 0.05 and K 0.04 mg g^{-1} soil. The soil was sterilized with formalin 15 days prior to seeding. Pots were seeded with sesame (cv Tilatara). In *Azospirillum* inoculated treatments prior to seeding, the seeds were surface sterilized by soaking in sodium hypochlorite solution (1 – 3% available chlorine) for 30 min at room temperature followed by washing with sterile distilled water. The seeds were then mixed in slurry containing lignite based *Azospirillum brasilense* (rice isolate) inoculant at 10^8 cells g^{-1} and dried in shade. In mycorrhizal treatments, 50 ml of *G. dimorphicum* inoculum containing chlamydospores (approximately 200 spores and infected root segment of *Sorghum* plants was uniformly mixed into the soil in all pots. The experiment consisted of four treatment (3 inoculated and 1 un-inoculated control) five graded levels of nitrogen and phosphorus (0, 25, 50, 75 and 100%) of the recommended dose (NP 30: 15 kgh^{-1}) and two stages of growth (35 and 70 days). All the 40 treatments were replicated three times. Three plants were sampled at 35 and 70 after emergence for monitoring various parameters.

2. 8. Identification of an efficient inoculation method for field grown sesame

2. 8. 1. Study site and soil characteristics

An experiment was conducted during 2001-2002 growing season on sandy (Entisol) soil to identify an efficient AM inoculation method for field grown sesame. Experimental areas of size 25m × 10m were chosen in a farmer's field at Avoor, Kerala. The field was ploughed using a power tiller and freed from the residue of the previous crop (rice). The area was equally divided into five blocks of 5 m × 10 m. Soil bunds of height 15cm were created between blocks to prevent water and soil movements. Random soil samples were withdrawn from the field to a depth of 30 cm to analyze major soil properties and indigenous AM spores. The soil characteristics before the start of the experiment was pH 5.2; organic carbon 1.40%; N 140 kgh^{-1}; P 21 kgh^{-1} and K 20 kgh^{-1}. Total indigenous AM fungal spore density in the soil was 150 spores per 100 ml soil consisting of *A. delicata, G. mosseae and G. versiformae.*

2. 8. 2. Crop and culture practices

Seeds of sesame (cv. Tilatara) were sworn at the rate of 4.5 kgh^{-1} in all plots followed by pressing with wooden plank so as to cover the seed in the soil. When the seedlings reached a height of 15 cm, thinning was done to avoid overcrowding. All treatments received N (urea) and P (muriate of potash) as broadcast at the rate of 30 kgh^{-1}. No P was applied to the soil in order to maximize the mycorrhizal benefit.

2. 8. 3. Experimental treatments

The experiment consisted of five inoculation treatment and one uninoculated control. In inoculated treatments, the inoculum of AM fungus *G. dimorphicum* was applied in the field in five methods (i) broadcast along with the seed, (ii) broadcast followed by land disturbance (iii) seed coating with AM fungi, (iv) gel entrapped AM fungi and (v) as mycorrhizal cakes (Fig.2.2).

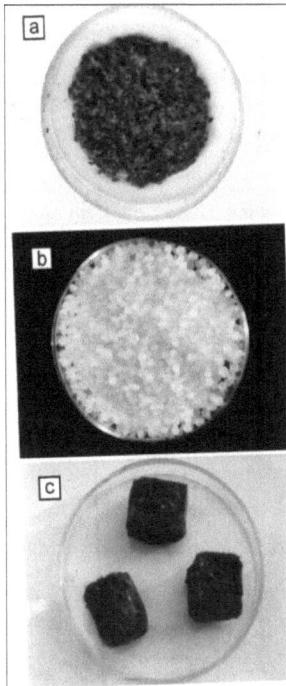

Fig 2. 2. Field inoculation method of AM fungi. (a) Sesame seeds coated with AM inoculums (b) Gel entrapped AM spores (c) Mycorrhizal cakes

In treatments received AM inoculation as broadcast the seeds were mixed with formaldehyde sterilized sand and inoculum (3:1) containing chlamydospores (approximately 20 spores g-1 soil) and infected root segment of Sorghum.

Prior to seed coating with AM fungi, both seeds and AM spores were surface sterilized by soaking in sodium hypochlorite solution (1 – 3%) available chlorine) for 30 min at room temperature. For sterilization of spores, the mycorrhizal inoculum was extracted from a pot culture of G. dimorphicum by wet sieving and decanting (Gerdemann and Nicolson, 1963). Sieving was collected on to the folds of a Whatman 40 filter paper kept inside a funnel. When water is drained off completely, the spores were flooded with a solution of 2 percent Chloramine T + 200 ppm Streptomycin sulphate + a drop of Tween 80 for 15 min. The spores were washed off to a beaker containing 100 ml lignite slurry. Fifteen ml of 1% aqueous solution of methylcellulose was added to it. The slurry was then mixed with 50g of sesame seeds and dried in shade. The coated seeds were broadcast in the respective plot.

Immobilization of AM spores was done by mixing 5.0 g sodium alginate with 100 ml distilled water containing the spore (surface sterilized) suspension. About 0.5 ml of sodium aliginate spore mixture was dropped in to 0.2 M $CaCl_2$. The spore entrapped alginate beads (diameter = 3.0+0.15 mm) formed by cross-linking with Ca^{2+} were harvested, washed with distilled water and used for field application.

Mycorrhizal cakes for field application were prepared by thoroughly mixing steam sterilized coconut oil cake and neem cake (5:1 ratio), surface sterilized sesame seeds and AM spore sieving in a 10% solution of pharmaceutical grade gum Arabic. Entire mix was moulded into 1 cm cubes and shade dried before application.

2. 8. 4. Sampling

Plants were sampled at 25, 50 and 75 days after emergence. Three randomly selected plants from each treatment were pulled out with the root system intact to evaluate various parameters.

2. 9. Analytical methods
2. 9. 1. Mycorrhizal colonization
2. 9. 1. 1. Assessment and quantification of AM fungi in plant roots

The undamaged fine roots of test plants were cut into 1 cm root segments from these about 30 segments were selected at random. The root segments were first washed thoroughly in distilled water and then placed separately in 10% KOH and heated to 90°C for 15 – 30 min. They were then washed in distilled water and immersed in alkaline 3% H_2O_2 for 5 – 10 min. The roots were again washed in distilled water and acidified with 5 N HCl for 2 – 5 min. Then the root segments were stained with 0.05% trypan blue in lactophenol for 15 – 30 min and the excess stain was removed with clear lactophenol (Phillips and Hayman, 1970). The root segments were mounted on clean microscopic slides in a mixture of glycerol and lactic acid (v/v). The root segments were gently squashed and covered by a glass cover slip and observed under a compound microscope (Nikon Eclipse E 400) using different magnification for AM fungal structures. Mycorrhizal colonization was assessed using the Trouvelot et al. (1986) method. The stained root sample was mounted in a mixture of glycerol and lactic acid (v/v). This method allows for the simultaneous evaluation of frequency of colonization (% F), intensity of colonization (%M), and the proportion of vesicles (%V) and arbuscules (%A) present in the roots. Frequency of colonization (%F) was calculated by the following equation.

$$F\% = 100 (N - no) /N$$

Where N is the number of root fragments observed and *no* is the number of root fragments without AM colonization. The intensity of AM colonization (%M) in each root segment was scored based in the whole presence of the fungus in the entire fragment using values from 0 to 5. Numbers indicate the proportion of root cortex colonized by the fungus, i.e. 0, without colonization; 1, colonization trace; 2, less than 10%; 3, from 11 to 50%; 4, from 51 to 90% and 5, more than 90% of the volume of root segment occupied by the fungus. Then the intensity of colonization (%M) as estimated by the following equation:

$$M\% = (95\ n5 + 70\ n4 + 30\ n3 + 5n\ 2 + n1) /N$$

Where n5, n4, n3, n2 and n1 are the numbers of fragments in the respective categories 5, 4, 3, 2, and 1. A similar system was used to calculate frequency of vesicles (%V) and arbuscules (%A), but in this case, the categories were 1. (without vesicles or arbuscules), 2. (less than 10%), 3. (from 11 to 50%) and 4, (more than 50% of the volume of the root fragment occupied by structures of these types) (Trouvelot et al., 1986).

2. 9. 1. 2. Spore analysis and extraction

Spores were extracted from a sub-sample (50 ml) of each soil sample by wet-sieving and sucrose density gradient centrifugation (Daniels and Skipper, 1982). Collected spores were washed in to a petridish and examined using a Zeiss Stemi-DV4 stereomicroscope. Healthy spores (based on shape, surface condition, spore content and no evidence of parasitism) of each distinct morphotypes (as distinguished by colour, shape and size) were directly counted. Spores of each morphotype are mounted on slides in polyvinyl alcohol-lactic acid-glycerol (PVLG) (Koske and Tessier, 1983) and PVLG mixed 1:1 (v/v) with Melzer's reagent. Spores were examined using a Nikon Eclipse E400 research microscope and identified up to species level using the manual for the identification of VA mycorrhizal fungi by Schenck and Pérez (1990) and comparing to voucher specimens and descriptions from the International Culture Collection of Arbuscular and Vesicular – Arbuscular Mycorrhizal Fungi (INVAM) (Morton et al., 1993). Spore density was determined as the number of healthy appearing spores per 50 ml soil.

2. 9. 1. 3. AM fungal diversity measurements

Species richness was defined as numbers of AM fungal species per soil sample. Isolation frequency was calculated as the percentage of samples in which the particular genus or species was present. Relative abundance was defined as the percentage of numbers of spores of the particular species or genera in the field soil.

2. 9. 2. Soil analyses
2. 9. 2. 1. Soil moisture

Soil sub samples 10.0 g were weighed into 50 ml glass beakers. These beakers were placed in a 105±1°C oven for 48 h, and the dry weight was then recorded. Gravimetric soil moisture was the difference in soil weights before and after oven drying (Gardner, 1986).

2. 9. 2. 2. pH

Soil pH was measured in a 1:2:5 fresh soil water suspensions with a digital pH meter (Systronics MK VI).

2. 9. 2. 3. Electrical conductivity

Electrical conductivity of the soil expressed as dSm^{-1} was determined in an aqueous extract (1:5) using a Systronics 304 conductivity meter (Piper, 1942).

2. 9. 2. 4. Organic carbon

Total organic carbon in soil was analyzed by dichromate oxidation and titration with ferrous ammonium sulphate (Walkley and Black, 1934).

2. 9. 2. 5. Available nitrogen, phosphorus and potassium

Available Nitrogen (N) in air-dried soil was determined by alkaline permanganate method (Subbaiah and Asija, 1956), Phosphorus (P) by ascorbic acid method (Wantanabe and Olsen, 1965) and exchangeable potassium (K) was extracted in an ammonium acetate solution and measured in Elico-CL345 digital flame photometer (Stanford and English, 1949).

2. 9. 3. Plant analyses
2. 9. 3. 1. Growth and yield components

Roots and shot length of the plants were measured using 1 ft scale. For taking the leaf number all the fully opened leaves from base to the top of the plant is counted, leaf area of about 10% of the total leaves from each plant was determined by outlining the leaves on graph paper and dry weight of these leaves was recorded. The leaf area per plant was computed using the leaf dry weight per plant and dry weight of those leaves for which the area was estimated (Watson, 1958). Yield components were determined at maturity.

2. 9. 3. 2. Dry matter
Total plant dry weight was recorded by drying the plants at 80°C till constant weight.

2. 9. 3. 3. Quality attributes
2. 9. 3. 3. 1. Oil content

Oil content in the seed was determined by Soxhlet extraction method. Five gram of dried sesame seeds was extracted with petroleum ether for 6 h in a Soxhlet system according to the AOCS method (AOCS, 1993). The oil extract was evaporated by distillation at a reduced pressure in a rotary evaporator at 40°C until the solvent was totally removed.

2. 9. 3. 4. Methods for biometric observation
2. 9. 3. 4. 1. Leaf area index

Leaf area index (LAI) was worked out by the formula suggested by Watson (1947)

$$LAI = \frac{\text{Leaf area of the plant (cm}^2)}{\text{Area of the pot}}$$

It is defined as the increase of plant material per unit of assimilatory material per unit time.

2. 9. 3. 4. 2. Net assimilation rate

Net assimilation rate (NAR) is defined as the increase of plant material per unit of assimilatory material per unit time. It was worked out using the following formula (Williams, 1946).

$$NAR = \frac{(W_2 - W_1)(\log_e L_2 - \log_e L_1)}{(t_2 - t_1)(L_2 - L_1)}$$

L_1 and W_1 are respectively the leaf area and dry weight of the plant at time t_1 and L_2 and W_2 are the leaf area and dry weight of the plant time t_2. This was expressed in g dm^{-2} day^{-1}.

2. 9. 3. 4. 3. Crop growth rate

Crop growth rate (CGR) of a unit area of a canopy cover at any instant in time (t) is defined as the increase in weight of plant material per unit of time. This was calculated by the formula.

$$CGR = NAR \times LAI \text{ (Watson, 1958)}$$

This was expressed in g m^{-2} day^{-1}

2. 9. 3. 4. 4. Specific leaf weight

Specific leaf weight (SLW) is the ratio of leaf dry weight to its area and is expressed in mg cm^{-2}

2. 9. 3. 4. 5. Harvest index

Harvest index (HI) is the ratio of seed weight to its total biological yield and is expressed as percent on dry weight basis.

$$HI = \frac{\text{Seed yield} \times 100}{\text{Total biological yield}}$$

2. 9. 3. 4. 6. Oil Index

Oil index (OI) was calculated using the following formula (Kohel, 1978)

$$\text{Oil index} = \frac{\text{Oil\%} \times 1000 \text{ seed wt}}{100}$$

expressed as g oil per 1000 seeds.

2. 9. 3. 4. 7. Specific P uptake

Specific P uptake (SPU) is expressed as total P uptake (mg P) per gram of dry root mass, and was calculated using the following equation (Zhu et al., 2003)

$$\text{SPU} = \frac{\text{Total P in plant (mg)}}{\text{Root dry weight (g)}}$$

2. 9. 3. 5. Biochemical
2. 9. 3. 5. 1. Tissue nutrients
2. 9. 3. 5. 1. 1. Nitrogen

Nitrogen (N) content in plant tissue was determined by the colorimetric method proposed by Snell and Snell (1949). Briefly, pulverized samples (0.5 g) taken in a digestion tube were soaked with 10 ml of a mixture of nitric acid, perchloric acid and hydrochloric acid (4:2:1) and kept overnight covering with a watch glass. It was then digested by gently heating on a hot plate until clear. The solution is washed out into a 50 ml standard flask and added 1ml of a 10% solution of sodium hydroxide and 1ml of 10% sodium silicate and diluted to volume with distilled water and mixed well. Five ml of the aliquot was pipette out in a 25 ml standard flask and added 1ml of Nessler's reagent and diluted to volume with distilled water. The colour intensity was read spectrophotometrically at 410 nm.

2. 9. 3. 5. 1. 2. Phosphorus

The Jackson (1973) method was used to determine the concentration of phosphorus (P) in plant tissues. Oven dried ground plant samples (0.5 g) were moistened with 10 ml nitric acid- perchloric acid sulphuric acid (10:4:1) mixture in a 50ml digestion tube overnight. The samples were digested using a hot plate until beaker containing 50ml of conc. H_2SO_4 at the bottom. The samples were digested using a hot place until white crystalline colour occurred to the powdered plant tissue. The contents were washed out into a 100 ml volumetric flask and diluted to volume with distilled water. Five ml aliquot of the diluted digest were taken in a 25 ml volumetric flask and added 5ml of vanadomolybdate colour reagent (containing 1 part, 25% nitric acid, 1 part, 0.13% ammonium metavanadate and 1 part, 2.5% ammonium molybdate) and kept it for 1-2 hrs to complete the reaction and diluted to volume with distilled water. Absorbance was read on a spectrophotometer (Systronics 104) at 470nm. The P content of plant tissues was calculated from tissue P concentration and tissue weight.

2. 9. 3. 5. 1. 3. Potassium

For the estimation of potassium (K), 5ml aliquot from diluted (100 ml) digests as prepared in item 2.9.3.3.1.2 was pipetted out into a 25 ml volumetric flask and diluted to volume with distilled water. The K content in the samples was measured by flame photometer (Elico CL 345) (Faithfull, 2002).

2. 9. 3. 5. 2. Total carbohydrate

Total carbohydrate in plant tissue was estimated as per the method of Hedge and Hofreiter (1962). Briefly, 1 g of plant tissue was weighed in to a boiling tube and hydrolyzed by keeping it in a boiling water bath for 3 hrs with 5ml of 2.5 NHCl and cooled to room temperature. It is then neutralized with solid sodium carbonate until effervescence ceases. The content of the tube was washed out into a 100 ml volumetric flask and made up to the volume. One ml aliquot taken from this was pipetted out into a test tube and added 4ml of ice-cold anthrone reagent and heated for 5 min in a boiling water bath. The content of the test tube was cooled rapidly and the intensity of green colour was read spectrophotometricaly at 630 nm.

2. 9. 3. 5. 3. Total soluble protein

Plant tissue weighing 0.5 g was ground in 10 ml of ice-cold. 1M-phosphate buffer (pH 7.0) using a pre- chilled mortar and pestle. The extract was filtered through cheese cloth and centrifuged at 10,000 rpm for 15 min. The supernatant was made up to a final volume of 100 ml using the extraction buffer. One ml aliquot from this was pipetted out into a test tube and added 5ml of Bradford reagent. After 3 min the absorbance was measured spectrophotometrically at 595 nm (Bradford, 1976).

2. 9. 3. 5. 4. Total free amino acids

Total free amino acids in plant tissue was estimated as per the method proposed by Yemm and Cocking (1955). Plant tissue (0.5 g) was ground using a mortar and pestle with a small quantity of acid washed sand. To this homogenate 5 ml of 80% ethanol was added and centrifuged. The extraction was repeated twice with the residue and all the supernatants were pooled. From this extract 0.1 ml was pipetted out into a test tube to which added 1ml of ninhydrin solution. The volume was made up to 2 ml with distilled water and heated the tubes in boiling water bath for 20min. Then 5 ml of the diluent mixture was added and mixed the content. After 15min, the intensity of purple colour was read spectrophotometrically at 570nm.

2. 9. 3. 5. 5. Proline

Free proline was estimated using the acid ninhydrin method (Bates et al., 1973). One gram of plant tissue was ground using a mortar and pestle with 6 ml of 3% (w/v) sulphosalicylic acid aqueous solution and the homogenate was filtered through Whatman No. 40 filter paper, then 2 ml of the filtered extract was taken for the analyses to which 2 ml of acid ninhydrin and 2 ml glacial acetic acid were added. The reaction mixture was incubated in a boiling water bath for 1 h and the reaction was finished in an ice bath. Four ml of toluene was added to the reaction mixture and the organic phase was extracted in which a toluene soluble reddish chromophore was obtained which was read at 520 nm using toluene as blank by Systronics 104 spectrophotometer.

2. 10. Statistical analysis

Analysis of variance (ANOVA) was used to determine the effect of each treatment and the interaction between them. In the case of variable which were expressed as percentages, data were transformed by the arcsine square-root procedure prior to ANOVA to ensure homogeneity of variable (Zar, 1999). For all characteristic studied the statistical significance of differences between means were determined using Turkey's HSD at P=0.05. Pearson correlation was used to study the relationship between different parameters studied. All statistical analyses were performed with MSTATC PROGRAMM (Michigan State University, East Lansing, Mich, USA).

CHAPTER 3

Arbuscular Mycorrhizal Associations with Field Grown Sesame

3. 1. Characteristics of sesame soils

Soil type of the sesame soils was predominantly sandy except in one location where laterite soil was observed. Physico-chemical characteristics of the soil varied significantly between locations (Table 3. 1). Soil moisture varied from 0.47% to 22.83%. In general, pH of the soil was acidic though it ranged from 4.3 to 6.5. The values of soil salinity ranged from 0.10 to 0.62 dSm^{-1}. Nevertheless, none of the sesame soils showed saline nature. Organic carbon levels of the soils ranged from 0.72 to 1.50%. Soil nutrients such as N, P and K also varied significantly in soils. Level of N ranged from 60.10 to 186.8 kgh^{-1} while P ranged from 26.7 to 230 kgh^{-1}. Soil K level ranged from 12.83 to 192.11 kgh^{-1}. The wide variation in soil physico-chemical characteristics observed in the present study could be attributable to the differences in cropping system and cultural practices received in each field (Christensen, 1988; Soon and Arshad, 1996; Agbenin and Goladi, 1997; Ju et al., 2007).

3. 2. AM colonization

Root colonization by AM fungi (Fig. 3.1) was highest in sample 18 (80%) and the lowest in sample 4 (5.8%). Out of the total number of samples only 22% had a comparatively high rate of colonization (>50%) while 78% of the samples had a low level of colonization (<50%). Harikumar and Potty (1999) in a similar study with sweet potato across field conditions, observed great variability in natural infection by AM fungi ranging from 25 to 100%. Observations of this nature have been described by other workers in crops such as cassava (Potty, 1978) and green gram (Valsalakumar et al., 2007).

Table 3.1. Soil physico-chemical characteristics and AM status in root and rhizosphere soils of sesame

No.	Soil Type	USDA Taxonomy	Soil characteristics								AM fungal status	
			Soil moisture (%)	pH	EC (dSm^{-1})	OC (%)	N	P Kgh^{-1}	K	Col. (F%)	Spore density (no. 50 ml^{-1} soil)
1	Laterite	Oxisol	0.47^i	5.4^{de}	0.49^b	1.50^a	143.7^d	200.0^{abc}	14.67^d	8.57^h	83^{ij}
2	Brown hydromorphic	Alfisol / Inceptisol	6.63^{fg}	5.2^{de}	0.39^{cd}	1.32^{bc}	146.3^d	230.0^a	21.33^d	7.33^h	158^{ef}
3	Brown hydromorphic	Alfisol / Inceptisol	11.10^{cd}	5.5^{cde}	0.28^{cdef}	1.27^c	186.8^a	140.0^{def}	26.5^d	20.03^g	138^{fg}
4	Greyish Onattukara	Entisol	4.30^h	5.5^{cde}	0.11^h	1.33^{abc}	112.3^{ef}	106.7^{fgh}	18.83^d	5.87^h	228^b
5	Greyish Onattukara	Entisol	11.86^c	5.0^{ef}	0.17^{gh}	1.26^c	148.9^{cd}	80.0^{hij}	17.17^d	17.00^{gh}	100^{hi}
6	Greyish Onattukara	Entisol	11.57^c	4.9^{efg}	0.36^c	1.36^{abc}	175.1^{abc}	70.0^{hij}	12.83^d	14.33^{gh}	195^{cd}
7	Greyish Onattukara	Entisol	11.90^c	5.3^{de}	0.10^h	1.42^{ab}	143.7^d	90.0^{gh}	17.17^d	24.83^{fg}	65j^k
8	Riverine alluvium	Entisol-Inceptisol	9.37^{de}	5.4^{de}	0.62^a	1.39^{abc}	185.5^{ab}	220.0^{ab}	21.33^d	14.07^{gh}	177^{de}
9	Greyish Onattukara	Entisol	17.3^b	5.2^{de}	0.14^h	1.27^c	138.2^{bc}	156.7^{cde}	19.50^d	8.67^h	210^{bc}
10	Greyish Onattukara	Entisol	8.73^e	4.4^{fg}	0.26^{defg}	1.30^{bc}	159.4^{bcd}	176.7^{bcd}	33.0^d	14.20^{gh}	215^{bc}

11	Greyish Onattukara	Entisol	8.90^e	4.3^g	0.28^{cde}	1.31^{bc}	182.9^{ab}	126.7^{efg}	17.83^d	6.83^h	272^a
12	Acid saline	Inceptisol	12.76^c	5.8^{bcd}	0.11^h	1.33^{abc}	109.7^f	100.0^{fgh}	27.17^d	34.13^{ef}	90^{hij}
13	Brown hydromorphic	Alfisol /Inceptisol	22.83^a	6.5^a	0.24^{efg}	0.96^e	112.3^{ef}	76.7^{hij}	139.60^{ab}	58.07^{bc}	143^f
14	Brown hydromorphic	Alfisol /Inceptisol	9.07^{de}	6.2^{ab}	0.20^{efgh}	0.92^e	60.10^h	230.0^a	76.0^c	47.63^{cd}	82^{ij}
15	Brown hydromorphic	Alfisol /Inceptisol	8.80^e	6.3^{ab}	0.19^{efgh}	1.02^d	91.48^{fg}	86.7^{ghi}	112.83^b	37.77^{de}	113^{gh}
16	Brown hydromorphic	Alfisol /Inceptisol	5.27^{gh}	6.1^{abc}	0.25^{efg}	0.92^e	75.78^{gh}	36.7^{jk}	16.00^d	66.66^b	72^{jk}
17	Brown hydromorphic	Alfisol /Inceptisol	7.43^{ef}	5.7^{bcd}	0.18^{fgh}	0.72^f	112.35^{ef}	43.3^{ijk}	19.50^d	62.23^b	53^h
18	Brown hydromorphic	Alfisol /Inceptisol	12.27^c	6.1^{abc}	0.25^{efg}	1.12^d	77.53^{gh}	26.7^k	192.17^a	80.00^a	112^{gh}

Means in each column with different letters are significantly different ($P<0.05$) by Tukey's HSD

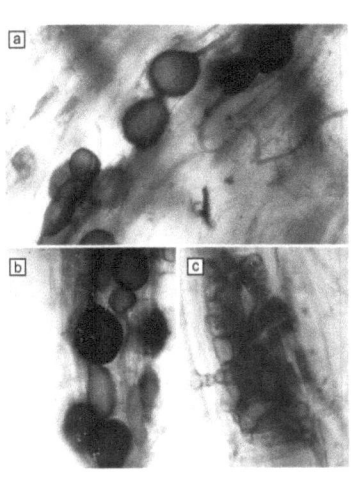

Fig. 3.1 AM colonisations in sesame; (a) AM structures observed within sesame roots collected from different sites (×10); (b) Vesicles in dense clusters (×10); (c) Arbuscules (×100)

The general inconsistency in AM colonization could be due to the influence of soil characteristics such as pH (Wang et al., 1993) organic carbon (Hršelová et al., 1999; Gryndler et al., 2002) and available nutrients (Johnson and Pfleger, 1992; Treseder, 2004; Schreimer and Linderman, 2005) as these variables exhibited significant relationship with root colonization by AM fungi (Table 3. 2). Genetic variability of the host plants grown in different fields may also be contributing to the differences in AM colonization in roots (Smith and Goodman, 1999). However, the present study did not make any lead in this direction.

3. 3. Spore density in soil

The mean value of spore density varied 5 fold among soil samples from 272 (sample 11) to 53 per 50 ml air-dried soil (Sample 17). The important factors likely to influence spore density are soil properties. AM fungi are sensitive to changes in soil pH as some species are restricted to either acid or alkaline soils; others in both (Porter et al., 1978; Robson and Abbot, 1989) which in turn reflects on spore density in soil. Here, within the soil pH range of 4.3 to 6.5, wide fluctuation in spore density was observed. This is probably due to the removal of species not adapted to a particular soil pH (Wang et al., 1993). Correlation study also supports this hypothesis as the spore density exhibited a significant negative relationship with soil pH (Table 3. 2). Soil organic carbon had a positive relationship with spore density in soil. Gryndler et al. (2002) observed that some of the organic materials in the soil could stimulate sporulation of AM fungi. Other soil factors such as N and P also exhibited a positive correlation with spore density. Soil K which is likely to be less significant with respect to AM fungi, exhibited an insignificant relationship with spore density. With a natural selection of species adapted to different soil conditions, it is not surprising that there was no limiting value found for soil nutrients to increase or decrease spore density. Thus the present results are inconsistent with the general hypothesis that spore density decreases with increasing nutrient status.

Table 3. 2. Pearson correlation between soil and AM fungal characteristics

	Moisture	pH	OC	EC	N	P	K	AM col. (%F)	Spore density (50 ml^{-1} soil)
Moisture	1.000								
PH	0.193	1.000							
OC	-0.138	-0.585***	1.000						
EC	-0.305*	-0.153	0.308*	1.000					
N	0.016	-0.718***	0.656***	0.445**	1.000				
P	-0.244	-0.265*	0.381***	0.486***	0.297*	1.000			
K	0.448***	0.607***	-0.444***	-0.133	-0.531***	-0.252	1.000		
AM col. (%F)	0.221	0.695***	-0.777***	-0.245	-0.715***	-0.579***	0.629***	1.000	
Spore density (50 ml^{-1} soil)	0.107	-0.579***	0.426***	0.166	0.525***	0.263*	-0.158	-0.596***	1.000

*P=0.05   *** P=0.001

Apart from soil factors, the spore density is influenced by the species of AM fungi infecting crop plants (Khalil and Loynachan, 1994). The previous crop may also have a major role in determining spore density in soil (Diaza et al., 1992). Since sesame is grown in rice fallows, the possible influence of the previous crop on the density and diversity of AM fungi cannot be ruled out. Nevertheless, these factors have not been examined in the present study.

3. 4. AM fungal diversity

Ten taxa of AM fungi were isolated from the soil samples belonging to the genera *Acaulospora, Gigaspora, Glomus* and *Scutellospora* (Table 3.3). Species richness in the survey area was 2.50. *Glomus* and *Acaulospora* were the dominant genera both in frequency and relative abundance (Table 3.4). The four most commonly observed species were *Glomus dimorphicum, Glomus versiformae, Acaulospora delicata,* and *Acaulospora lacunosa.* Of these *Glomus dimorphicum* was the most frequent and abundant species (Fig. 3. 2 – 3. 4).

AM fungi are known to exhibit ecological specificity (McGonigle and Fitter, 1990). There is no published report on the AM fungal community associated with sesame under the climatic conditions prevailing on the sesame soils of Kerala.

Table 3.3. Distribution of AM species in rhizosphere soil samples of sesame

AM fungi	Soil samples																	
	1	2	3	4	5	6	7	8	9	10	11	12	13	14	15	16	17	18
Acaulospora delicata Walker, Pfeiffer & Bloss	+	+				+	+	+										+
Acaulospora lacunosa Morton	+	+	+															+
Gigaspora decipiens Hall & Abbot														+		+	+	
Gigaspora margarita Becker & Hall												+				+		
Glomus aggregatum Schenck & Smith emend. Koske												+						
Glomus dimorphicum Boyetchko & Tewari	+	+	+	+	+	+	+		+	+	+	+	+					
Glomus fasciculatum (Thaxter) Gerdemann & Trappe emend Walker & Koske													+		+		+	
Glomus mosseae (Nicolson & Gerdemann) Gerdemann & Trappe								+								+		
Glomus versiformae (Karsten) Berch		+	+	+	+				+	+	+		+	+	+	+		
Scutellospora nigra (Reddeard) Walker & Sanders																		+
No. of species sample^{-1}	3	4	3	2	2	2	2	2	2	2	2	3	3	2	2	4	2	3

+ = present

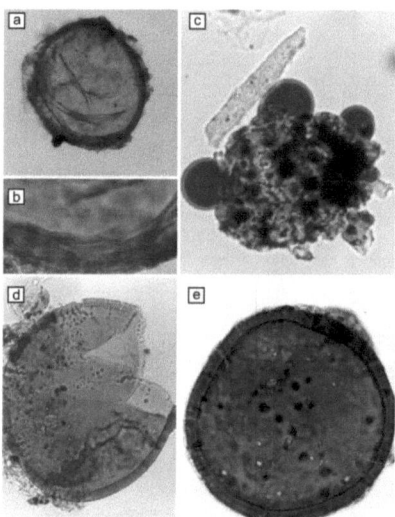

Fig. 3.2. a-e. Photomicrographs of AM fungal isolates from soil samples : (a) *Acaulospora delicata* (×40) ; (b) Spore of *A. delicata* showing wall thickness (×100) ; (c) *Acaulospora lacunosa* (×40) ; (d) Immature spore of *Acaulospora lacunosa* (×40) ; (e) Mature spore of *Acaulospora lacunosa* (×40)

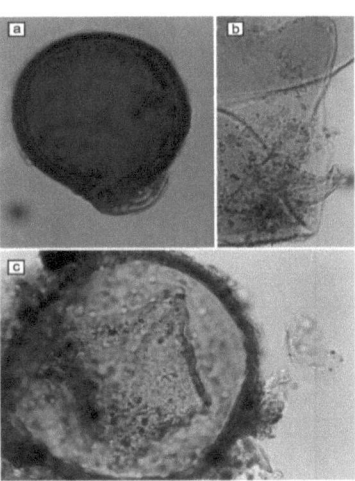

Fig. 3.3. a-c. Photomicrographs of more AM fungal isolates from soil samples ; (a) *Glomus dimorphicum* (×40) ; (b) Broken spore wall of *G. dimorphicum* showing ornamentation (×100) ; (c) *Glomus versiformae* (×40)

Fig. 3.4. a-b Photomicrographs of more AM fungal isolates from soil samples ; (a) *Scutellospora nigra* (×40); (b) Spore wall ornamentation in *S. nigra* (×100)

Table 3. 4. Isolation frequency (*F*, %) and relative abundance (RA%) of the four AM fungal genera isolation sesame

Genus	F	RA
Acaulospora	55.55	23.91
Gigaspora	16.66	4.36
Glomus	94.44	69.77
Scutellospora	5.55	1.88

Only a few studies have addressed the incidence of mycorrhizal association in the roots of sesame that too in wild species (John et al., 2007). The predominance of *Glomus* and *Acaulospora* is in accordance with Klironomos et al. (2001), who studied interspecific difference in AM fungal tolerance to freezing and dry conditions. These authors found that *Glomus* and *Acaulospora* were most frequently isolated in the field and *Glomus* species were the least affected by freezing in pot conditions while drying gave more variable responses in colonization by AM fungi. Soil properties are important factors influencing the AM fungal community (Bever et al., 2001). There are contradictory findings on the distribution of *Glomus* species in relation to soil pH. For example, Mosse (1973) reported that some *Glomus* species are very common in neutral or alkaline soil but few in acid soils, whereas species of *Acaulospora* are usually found in acid soils. In contrast to this the predominance of *Glomus* species in acidic soils with a pH ranging from 4.2 to 6.8 has been reported from South India (Dalal and Hippalgaonkar, 1995; Bhardwaj et al., 1997; Valsalakumar et al., 2007). In the present study, though all the soils remained within the acidic range, *Glomus* appears to dominate indicating its tolerance to soil acidity and this could be one explanation for frequent detection of *Glomus*. Soil type and organic matter content are also likely to influence AM fungal community composition.

3. 5. AM fungi and nutrient range

Nutrient range for AM fungi associated with sesame is presented in Table 3. 5. *Glomus* and *Acaulospora* were found to occur in a wider nutrient range than that of *Gigaspora* and *Scutellospora*. Among the *Glomus* species, *G. dimorphicum* was found to occur in the maximum range. The wide range of nutrient tolerance could be one of the possible reasons for the wide distribution of *Glomus* in sesame soils. Nevertheless, the results of correlation studies indicate that AM fungal colonization and sporulation are not mearly controlled by soil nutrients as was suggested by Hayman (1970).

Table 3. 5. Isolation frequency (*F %*) and relative abundance (RA%) of AM fungal species isolated from sesame soils

Species	F (%)	RA
A. delicata	38.88	16.73
A. lacunosa	27.77	7.22
G. decipiens	11.11	2.5
G. margarita	5.55	1.85
G. aggregatum	5.55	1.53
G. dimorphicum	72.22	40.26
G. fasciculatum	11.11	4.77
G. mosseae	11.11	4.82
G. versiformae	61.11	18.36
S. nigra	5.55	1.88

F Isolation frequency; RA relative abundance

Table 3. 6. Soil nutrient range and AM fungal species in sesame soils

	N	P	K
	kgh^{-1} (range)		
A. delicata	77.53-185.50	26.70-230.00	12.83-192.17
A. lacunosa	60.10-186.80	36.7-230.00	14.67-76.00
G. decipiens	75.78-77.53	26.7-36.70	16.00-192.17
G. margarita	109.70	100	27.17
G. aggregatum	109.70	100	27.17
G. dimorphicum	91.48-186.80	70-230	14.67-139.60
G. fasciculatum	112.30	43.30-76	19.50-139.60
G. mosseae	75.78-185.50	36.70-220.00	16.00-21.33
G. versiformae	60.10-186.80	80-230	16.00-139.60
S. nigra	77.53	26.70	192.17

3. 6. Conclusions

The field study on AM fungal association in sesame in Kerala revealed that there exists wide variability in root colonization and spore density in rhizophere soil samples. Soil characteristics had a profound influence on both root colonization and spore density in soil. The genus *Glomus* with its predominant species *G. dimorphicum* was the dominant one both in frequency and relative abundance. Since this species is the frequent associate with the crop under study, the efficiency of this species has to be evaluated in pots as well or in field. The results of such studies are presented in the subsequent chapters.

CHAPTER 4

Development of Fungal Structures by Indigenous AM Fungi during the Growth of Sesame

Sesame plants inoculated with all the five indigenous AM fungi proved its infectivity by producing intercellular hyphae (*Arum*- type) vesicles and arbuscules within the roots indicative of a functional AM association in this crop. The fungal parameters such as frequency of colonization (%F), intensity of hyphal colonization (%M), frequency of vesicles (%V) and arbuscules (%A) varied significantly (Table 4. 1) in plants received inoculation with different AM fungi.

Table 4. 1. ANOVA of main effects and their interaction on the development of fungal structures in sesame inoculated with different indigenous AM fungi

Source	df	%F		%M		%V		%A	
		MQ	P	MQ	P	MQ	P	MQ	P
AM fungi (1)	4	1871.96	0.000	3.22.39	0.000	4803.21	0.000	682.74	0.000
Plant age (2)	6	5970.99	0.000	936.45	0.000	1422.70	0.000	299.29	0.000
1×2	24	621.91	0.000	122.207	0.000	754.05	0.000	271.21	0.000
Error	70	121.34		33.56		56.43		6.91	

However, more vigorous development of fungal structures was observed in *G. versiformae* inoculated plants and the least in *A. delicata*, A. *lacunosa* and S. *nigra* inoculated ones (Fig 4. 1). This could be possibly due to the differences in symbiotic effectiveness, which depends on the preference of a particular AM species to a definite host (Dhillion, 1992). The existence of similar variability in the development of AM fungal structures in some plant species of coastal sand dunes of Paraguana' Peninsula, Venezuela has been reported by Alarcón and Cuenca (2005).

Development of AM fungal structures was generally low during the initial stages of plant growth but subsequently increased to maximum during the period of vegetative growth (Sutton, 1973; Reddy et al., 1997). Typically a higher degree of AM colonization coincides with the stages of plant life cycle that demand additional P (Rabatin, 1979; Dhillion et al., 1988; Sanders and Fitter, 1992; Anderson et al., 1994). Therefore, it is reasonable to assume that the high rate of development of fungal structures could be linked with the increasing demand of P during the period of vigorous plant growth. Hayman (1983) reported that the extent of formation of fungal structures such as vesicles and arbuscules is known to influence nutrient exchange process.

In the present study, the development of all the fungal structures declined during the yield stage (Fig 4.1) similar to the results reported in wheat by Cade-Menun et al. (1991). They suggested that as the grain ripens; photosynthesis slows down and the nutrients are translocated from the leaves to the grain (Karlen and Whitney, 1980). Therefore, the level of photosynthate supply to the roots may correspond to the decline in colonization observed during the yield stage. Other researchers (Jakobsen and Nielson, 1983) have also found that AM levels off during the late stages of wheat growth. Progressive death of fungal structures with the advancement of plant age could be another possible reason for this decline (Smith and Read, 1997).

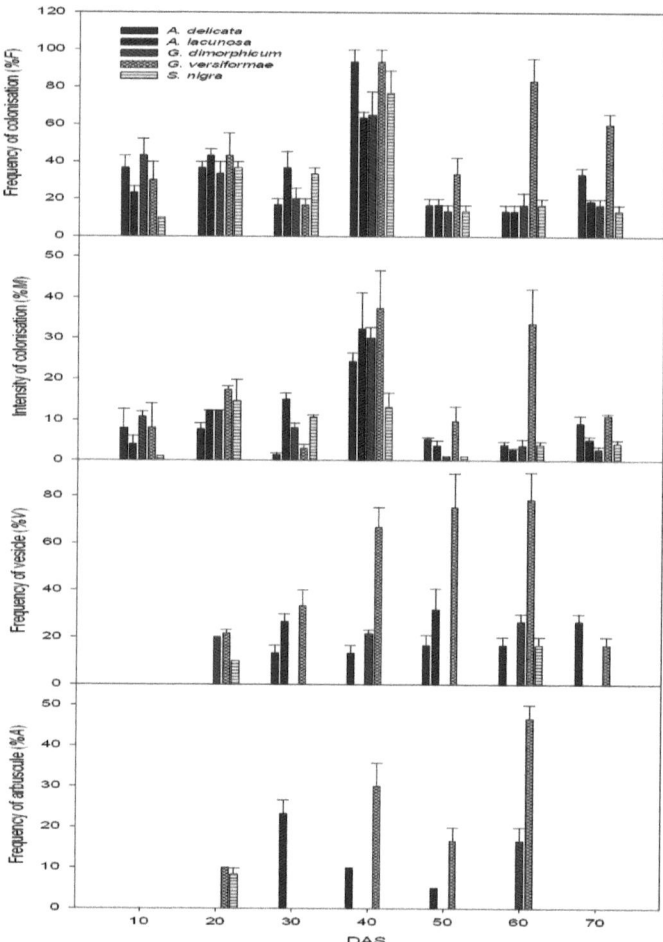

Fig. 4. 1. Changes in the developmental pattern of AM fungal structures during the growth of sesame inoculated with different indigenous AM fungi.

Pearson correlation analysis revealed that there existed relationship between different AM fungal parameters (Fig 4. 2). % F showed significant positive relationship with % M (r = 0.832, df 10, P<0.001), %V (r=0.367, df 10, P<0.001) and %A (r = 0.472 df 10, P<0.001).

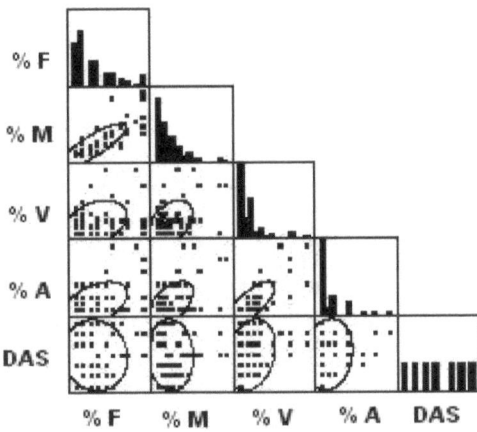

Fig. 4. 2. Scatter plot matrix showing the relationship between fungal variables.

Similarly % M showed a significant linear relationship with %V (r = 0.436, df 10, P<0.001) and % A (r = 0.539, df 10, P < 0.001). %F and %M however showed an insignificant inverse relationship with plant age (DAS). Among the other parameters (%V and %A), %V alone showed a significant (r = 0.232, df 10, P<0.017) positive relationship with plant age. A number of workers have described the relationship between root colonization by AM fungi and spore density in soil (Li et al., 2007; Zak, 2009). But they failed to find the relationship between the fungal parameters examined in the present study.

4. 1. Conclusion

The study provides an unequivocal evidence of AM association in sesame. The developmental pattern of fungal structures varied with inoculants with a more vigorous development of fungal structures in *G. versiformae* inoculated plants. Irrespective of inoculants, the development of fungal structures reached the maximum level at vegetative growth of the host indicating the importance of AM symbiosis during the period of rapid vegetative growth. The relationship between the fungal parameters warrants the need of scoring all these variables while evaluating the efficiency of mycorrhizal system.

CHAPTER 5

Relative Efficiency of Different Indigenous AM Fungi on the Growth and Yield of Sesame

5.1. Morphological characteristics

Morphological characters like rootlet number, shoot length, leaf number and leaf area were significantly (all $P < 0.007$) influenced by inoculation with indigenous AM fungi (Table 5.1). However, the isolates varied in their capacity in enhancing these parameters. The highest rootlet number and shoot length was observed in *A. lacunosa* inoculated plants while *G. dimorphicum* inoculated plants had the highest leaf number and leaf area. Further, morphological characters showed significant ($P < 0.001$) difference with DAS. In general, the highest values for these characters were observed at harvest (75 DAS) (Fig. 5.1-2). The influence of AM fungi on increased plant growth is perhaps due to increased P uptake which might have caused cell multiplication and elongation (Black, 1965). However, there existed variation in their effectiveness, which could be due to the differences in the uptake of P and other nutrient elements in plants inoculated with different fungi (Rakshit and Bhadoria, 2008). This differences may be attributed to (1) differences among AM fungi for hyphal spread and density away from roots (Bürkert and Robson, 1994), (2) ability of AM fungi to increase nutrient availability, especially P, in soil through enhanced phosphatase/phytase activity (Dinkelaker and Marschner, 1992; Khalil et al., 1994) and/or excretion of solubilizing materials such as ethylene (Ishii et al., 1996), flavonoides (Ishii et al., 1997), and growth regulating compounds (Danneberg et al., 1992; Thiagarajan and Ahmad, 1994), and (3) ability of AM fungi to change rhizosphere soil pH (Gianinazzi-Pearson and Azcón–Aguilar 1991; Li et al., 1991).

Table 5. 1. Effect of inoculation with indigenous AM fungi on the growth of sesame

Treatment	Rootlet no. plant^{-1}	Shoot length (cm)	Leaf no. plant^{-1}	Leaf area (cm^{-2} plant^{-1})
Control	24.00^b	23.17^b	9.79^c	39.96^c
A. delicata	31.33^{ab}	25.79^{ab}	11.22^{bc}	49.26^{bc}
A. lacunosa	37.00^a	28.67^a	12.44^{abc}	84.30^a
G. dimorphicum	33.00^a	29.50^a	14.89^a	88.42^a
G. versiformae	30.00^{ab}	25.71^{ab}	13.33^{ab}	71.59^{ab}
S. nigra	31.44^{ab}	26.29^{ab}	14.67^a	66.26^{ab}

Means in each column with different letters are significantly different ($P<0.05$) by Tukey's HSD

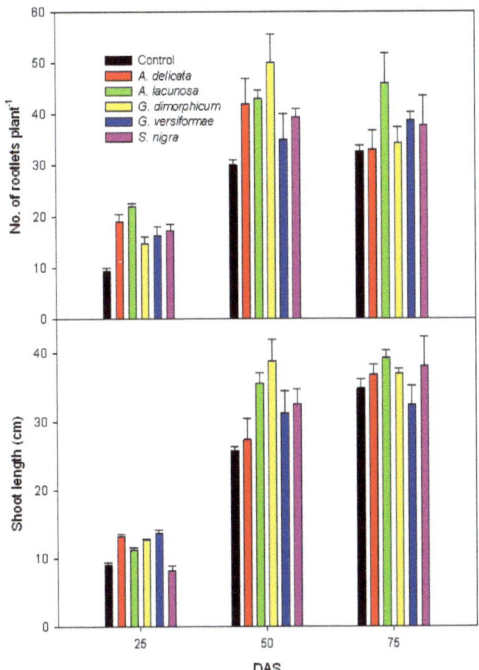

Fig. 5.1. Effect of inoculation with indigenous AM fungi on rootlet number and shoot length at different stages of crop growth of sesame

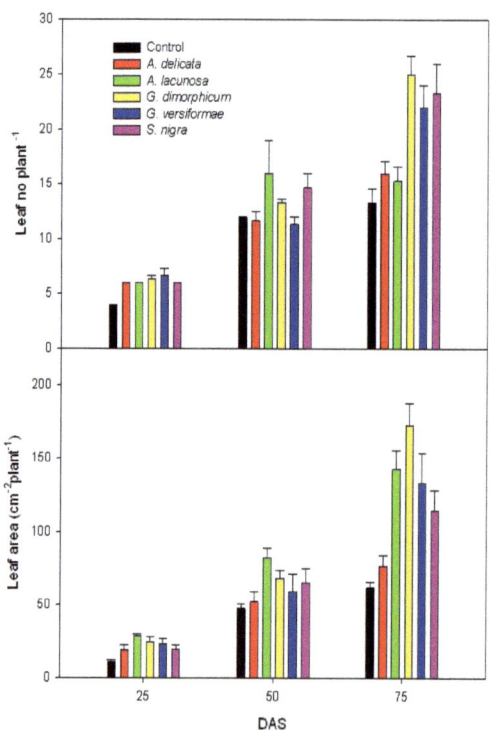

Fig. 5.2. Effect of inoculation with indigenous AM fungi on leaf number and leaf area at different stages of crop growth of sesame

Similar differences in the performance of different species of AM fungi as in the present study have been reported in crops such as *Paspalum notatum* (Mosse, 1972) and sugarcane (Reddy et al., 2004).

5. 2. Biomass production

Inoculation with indigenous AM fungi markedly increased biomass in sesame plants over uninoculated control (Table 5. 2). In inoculated treatments, the fresh and dry weight ranged from 1.95 to 2.76 and 0.34 to 0.56 g respectively. The plant fresh weight increased 65.26% over control in *A. lacunosa* inoculated treatment, while the plant dry weight increase two fold. Plant biomass varied significantly ($P < 0.001$) with plant age, which reached a maximum increase at 75 DAS (Fig. 5. 3) Declerck et al. (1995) investigated the growth response of micro-propagated banana plants to AM inoculation.

Table 5. 2. Effect of inoculation with indigenous AM fungi on biomass production in sesame

Treatment	Plant biomass (g)	
	Fresh	Dry
Control	1.67^b	0.28^c
A. delicata	2.09^{ab}	0.38^{bc}
A. lacunosa	2.76^a	0.56^a
G. dimorphicum	2.41^{ab}	0.40^b
G. versiformae	1.95^{ab}	0.34^{bc}
S. nigra	1.99^{ab}	0.52^a

Means in each column with different letters are significantly different ($P<0.05$) by Tukey's HSD

The authors report that inoculation with *Glomus mosseae* and *Glomus geosporum* resulted in a significantly higher shoot and root dry weights as compared to the control plants.

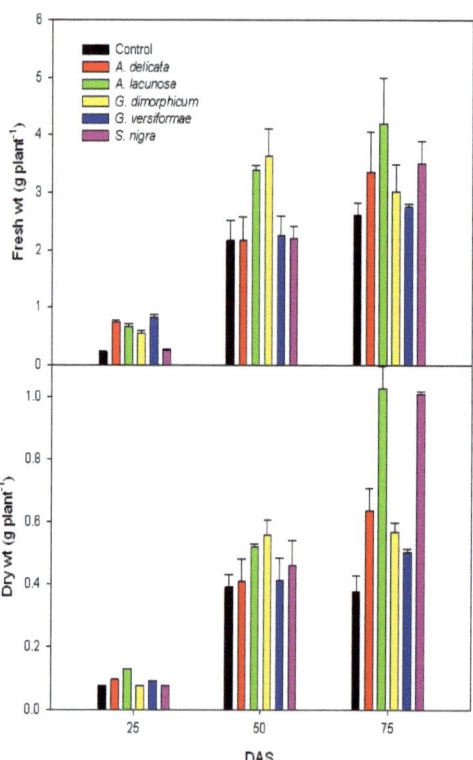

Fig. 5.3. Effect of inoculation with indigenous AM fungi on plant biomass (fresh and dry) at different stages of crop growth of sesame

Fortuna et al. (1992) observed large differences in the fresh and dry mass between inoculated and un-inoculated plum plants as a result of differences in the growth behavior of the plants. According to Branzanti et al. (1992) and Azcón-Aguilar and Barea (1997) mycorhiza enhances growth of plantlets of selected species and cause earlier resumption in shoot apical growth. Vestberg (1992) found that only 3 of 6 fungal strains tested with 10 strawberry cultivars were highly efficient with regard to significant growth improvements.

5. 3. Yield components

Yield components such as pod number, pod weight and seed number were significantly ($P < 0.035$) enhanced in treatments inoculated with indigenous AM fungi (Table 5. 3). However, in the case of seed weight, the increase has not reached a significant level. Among the various AM fungi tested, inoculation with *G. dimorphicum* markedly increased the yield components in sesame.

Table 5. 3. Effect of inoculation with indigenous AM fungi on yield components of sesame

Treatment	Pod no plant^{-1}	Pod wt. (g plant^{-1})	Seed no plant^{-1}	Seed wt. (g plant^{-1})
Control	0.33^c	0.18^b	9.59^b	0.33^a
A. delicata	0.89^{ab}	0.20^{ab}	10.33^{ab}	0.36^a
A. lacunosa	0.78^{ab}	0.20^{ab}	12.67^{ab}	0.47^a
G. dimorphicum	1.11^a	0.35^a	21.56^a	0.65^a
G. versiformae	0.89^{ab}	0.25^{ab}	17.22^{ab}	0.52^a
S. nigra	0.67^{bc}	0.24^{ab}	20.77^{ab}	0.62^a

Means in each column with different letters are significantly different ($P<0.05$) by Tukey's HSD

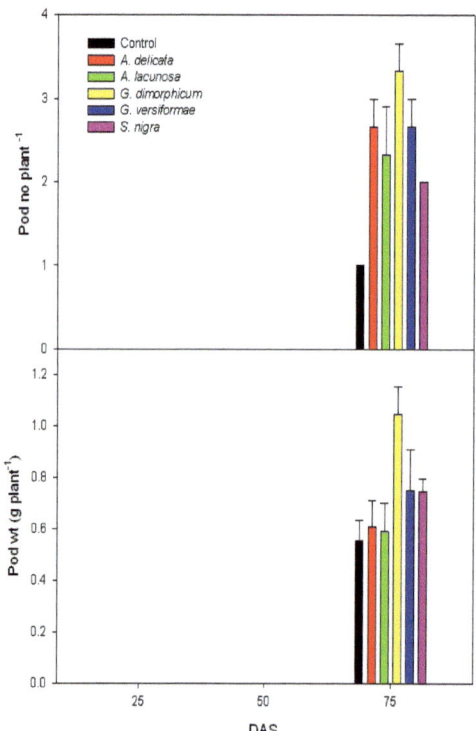

Fig. 5. 4. Effect of inoculation with indigenous AM fungi on yield components (pod number and weight) at harvest stage of sesame

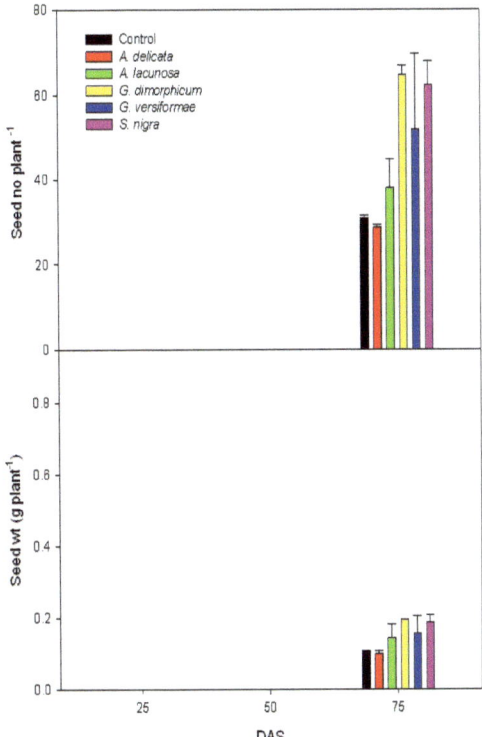

Fig. 5.5. Effect of inoculation with indigenous AM fungi on yield components (seed number and weight) at harvest stage of sesame

Since the reproductive stage of the crop starts at 50 DAS, the yield components could be gauged only at 75 DAS (Fig. 5. 4-5). Increased yield consequential to AM inoculation has been reported in crops such as coffee (Siqueira et al., 1998), barley (Khaliq and Sanders, 2000) and *Trifolium alexandrium* (Shokri and Maadi, 2009).

5. 4. AM colonization

Mycorrhizal colonization (%F) was significantly ($P < 0.001$) higher in all the treatments inoculated with indigenous AM fungi (Fig 5. 6).

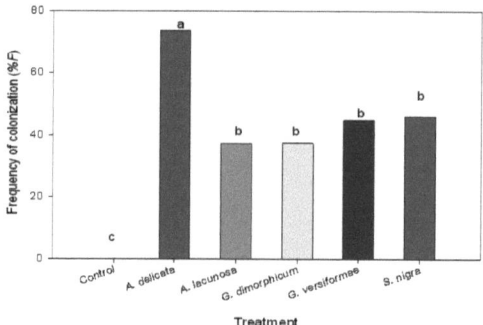

Fig. 5. 6. Effect of inoculation with different indigenous AM fungi on mycorrhizal colonization in sesame. Bars with different letters are significantly different ($P<0.05$) by Tukey's HSD

Different isolates colonized sesame roots to different levels ranging from 37.32 to 73.67%. The highest %F was observed in *A. delicata* inoculated plants. Irrespective of AM inoculant, the % F was highest at 50 DAS (Fig. 5. 7). It was observed that the beneficial effect from a particular species of AM fungi was not always correlated with the extent of root infection.

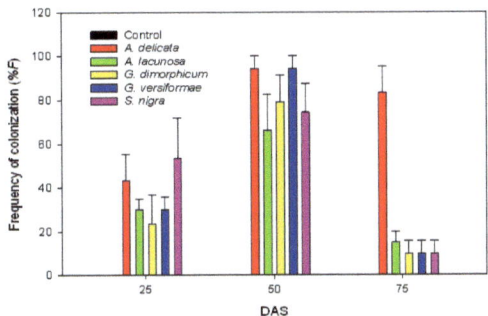

Fig. 5.7 Effect of inoculation with indigenous AM fungi on mycorrhizal colonization (%F) at different stages of crop growth of sesame

For example, sesame inoculated with *G. dimorphicum* had only 37.44% of its root colonized by AM fungi but maximum increase in 63% of the measured parameters was recorded with this fungus. As has been observed elsewhere, AM fungi differ in their ability to enhance growth of the host plant, regardless of the extent of root colonization (Graham et al., 1982). One of the most important factors that influence the efficiency of different AM fungal strains seems to be their external mycelium. The production of external hyphae may vary considerably between AM fungi (Sanders et al., 1977; Abbot and Robson, 1985; Kothari et al., 1991). No clear relationship seems to exist between the amount of external hyphae in soil and the growth responses observed in colonized plants (Jakobsen et al., 1992; Frey and Schuepps, 1993). Other factors such as the difference in rate of appresorium formation, in hyphal uptake and translocation capacities of nutrients, and in the metabolic activity of the external hyphae, seem to have more influence on the efficiency of AM fungi (Jakobsen et al., 1992, Frey and Schuepps, 1993; Giovannetti and Citernesi, 1993).

5. 5. Conclusion

In general, the indigenous AM fungi improved the growth and yield characters of sesame though their efficiency varied. Among the AM fungi, *G. dimorphicum* emerged out as the efficient isolate in improving majority of the tested parameters. The study thus sheds light into the importance of proper selection of efficient AM fungi for the right crop and environment.

CHAPTER 6

Infectivity of Spores of AM Fungus *Glomus dimorphicum* on Sesame Plants after Long Term Incubation in Different Soil Types under Two P Regimes

The infectivity of AM fungus *G. dimorphicum* on sesame plants varied significantly ($P<0.001$) after long-term incubation of spores in soil types with and without P addition. Frequency of AM colonization (%F) offered to plants by spores incubated in LA soil tended to be larger than those incubated in other soil types (Table 6. 1).

Table 6. 1. Infectivity of *G. dimorphicum* on sesame plants after long-term incubation in different soil types under two P regimes

Soil type	Soil treatment			
	P0		P1	
	AM colonization			
	Frequency (%F)	Intensity (%M)	Frequency (%F)	Intensity (%M)
CA	28.98^{cd}	3.57^b	22.72^e	3.34^b
GO	24.61^{de}	3.60^b	16.44^f	4.19^b
KA	35.83^b	8.38^a	7.22^g	1.03^b
LA	47.11^a	7.96^a	33.16^{bc}	3.27^b

Means in each column with different letters are significantly different ($P<0.05$) by Tukey's HSD

Addition of P to soils significantly ($P<0.001$) reduced the %F in plant roots irrespective of spore incubation in different soil types. The %F was also significantly ($P<0.001$) affected by duration of spore incubation in soil. In general, the %F recorded a high value in roots during early period of spore incubation in soils and declined greatly after 5 months after incubation (MAI) (Fig. 6. 1).

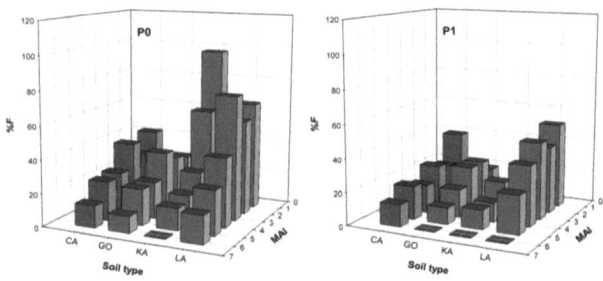

Fig. 6. 1. Frequency of AM colonisation (%F) in roots of sesame after long-term incubation of spores of *G. dimorphicum* in different soil types under two P regimes.

Intensity of colonization (%M) in sesame roots like %F varied significantly ($P<0.005$) after long-term incubation of spores in soil types with and without P addition. Higher %M was offered to the roots of test plants by spores of *G. dimorphicum* incubated in KA and LA soils during infectivity tests. However, addition of R to these soils significantly ($P<0.001$) reduced the %M in roots harvested from these soil types. Significant ($P<0.001$) differences in %M in roots were observed with a period of spore incubation. Higher %M was observed in plant roots got an infection during the early period of spore incubation up to a maximum of 5 MAI and declined thereafter (Fig. 6. 2).

The results showed that long-term incubation of spores of *G. dimorphicum* in different soil types under varied P regimes greatly affected both spore viability in soils and its infectivity in sesame plants.

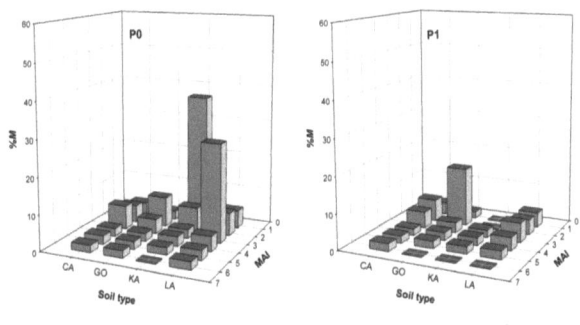

Fig. 6. 2. Intensity of AM colonisation (*%M*) in roots of sesame after long-term incubation of spores of *G. dimorphicum* in different soil types under two P regimes.

Land and Schöönbeck (1991) reported that soil type has a profound influence on spore density and root colonization by AM fungi. The infectivity of the fungus on host expressed in terms of *%F* and *%M* was comparatively high under spore incubation in laterite soil. This contrasts the finding of Russel (2007) that sandy soils are more conducive to mycorrhiza formation than clays or peat bogs because of their non porous, warmer, drier and less fertile nature (Sylvia and Williams, 1992). P fixation has been reported as a phenomenon limiting P availability in acid laterite soils of the tropics (Tandon, 1987). The rapid P fixation taking place in this soil might have removed P in its bound form lowering its deleterious effects on spore viability and infectivity (Hayman, 1970; Giovannetti, 2000). This could be a possible explanation for a high colonization rate in plant roots harvested from this soil type. However offering such comments need more in depth studies.

Application of P as rock phosphate into soil impeded root infectivity by AM fungi in most soils types (Mosse, 1973). However, in the case of plant roots harvested from LA soil, there was no apparent decrease in root colonization even after P addition to soil. As mentioned earlier, the fast rate of P fixation offsetting the adverse effect of P on spore viability and infectivity might be the reason for this.

Long-term incubation of spores in all soil types decreased its infectivity on sesame irrespective of P fertilization. However, the infectivity of spores was greatly affected after 5 MAI which is in fine with the finding of Ruiz-Lozano and Azcón (1996) that the infective capacity of AM spores decreased considerably after 6 months of incubation in soils with different water potentials. The decrease in infectivity observed in the present study with increasing incubation time could be ascribable to loosing of viability or complete death of propagates in the absence of host (Logi et al., 1998; Giovannetti, 2004).

6. 1. Conclusion

Viability and infectivity of spores of *G. dimorphicum* varied after incubation in different soil types. Addition of P to soil negatively affected the viability and infectivity of AM fungi most soils except in LA. Though the viability and infectivity of the spores were affected negatively after long-term incubation in soils, the spores retained viability and infectivity up to 5 months in most soil types. The study, therefore warrants the need of an AM inoculation strategy if the fallow period is much prolonged before sesame cultivation.

CHAPTER 7

Variability in natural infection by AM fungi and its influence on the growth and P nutrition of sesame accessions

Sesame accessions exhibited variation in both frequency (%F) and (%M) intensity of root colonization by AM fungi. Among the 20 accessions screened, 30% had a comparatively higher (>40%) frequency of colonization whereas; in other ones the frequency was more or less on par. Intensity of colonization was comparatively more (>8%) in 40% of the accessions while two accessions (SI, 68, 69) had a very low intensity of colonization (Table 7. 1). While AM fungi are known to vary in their ability to colonize and transfer P to the plant and confer other beneficial effects, little is known to the exact role of the host genotype in the expression of AM fungi. In the present study, though the accessions were grown in the same field under identical environmental condition, there was significant variation in the frequency and intensity of root colonization by AM fungi which could be ascribed at least in part to genetic and physiological factors controlling host/fungus compatibility. Recently discovered mycorrhizal mutants (myc⁻) of pea offer great promise for the study in this direction (Graham and Eissenstat, 1994). Similar genotype dependent variation for AM fungal colonization has been reported in wheat (Azcón and Ocampo, 1980) sorghum (Clark, 1983) pearlmillet (Krishna et al., 1985) coconut (Thomas and Ghai, 1987) and cowpea (Mercy et al., 1990).

Correlation studies between fungal characters viz. frequency of colonization and intensity of colonization revealed that there exists a significant positive relationship between these variables. The similar significant positive correlation was observed between fungal and plant characters (Table 7. 2).

Table 7. 1. Frequency of colonization (%F) and intensity of colonization (%M) in sesame accessions

Accession no.	Colonization	
	%F	%M
SI 2	43.33 ^{bcd}	8.70^c
SI 7	43.33^{bcd}	9.17
SI I5	26.67^{ef}	5.66^c
SI I7	23.33^{ef}	4.50^{ef}
SI 25	26.67 ^{ef}	5.90^{de}
SI 30	56.67^{ab}	17.70^a
SI 32	46.67^{bc}	8.17^{cd}
SI 37	28.33^{def}	4.00^{ef}
SI 42	28.33 ^{def}	4.93^{ef}
SI 47	68.33^a	19.30^a
SI 48	50.00^b	14.50 ^{b}
SI 56	33.33^{cde}	8.90^c
SI 57	33.33^{cde}	8.60^c
SI 58	26.67^{ef}	5.43^{ef}
SI 59	26.00^{ef}	5.27^{ef}
SI 63	23.33^{ef}	4.50^{ef}
SI 66	23.33^{ef}	5.07^{ef}
SI 68	16.67^f	3.07^f
SI 69	28.33^{def}	3.63^{ef}
SI 70	28.33^{def}	4.13^{ef}

Means in each column with different letters are significantly different ($P<0.05$) by Tukey's HSD

Table 7. 2. Pearson correlation between AM fungal and plant characters in sesame

Parameters	Mycorrhizal parameters	
	% F	*% M*
Mycorrhizal parameters		
% F	1.000	
% M	0.897***	1.000
Growth characters		
Root length (cm)	0.479***	0.547***
Shoot length (cm)	0.499***	0.528***
Leaf Number plant $^{-1}$	0.707***	0.781***
Biomass production		
Root fresh weight (g)	0.593***	0.669***
Shoot fresh weight (g)	0.493***	0.563***
Total biomass (g)	0.538***	0.614***
Dry matter production		
Root dry weight (g)	0.573***	0.667***
Shoot dry weight (g)	0.526***	0.621***
Total dry weight (g)	0.544***	0.641***
Yield		
Pod Number (plant^{-1})	0.404***	0.358***
P efficiency		
P content (mg g^{-1})	0.546***	0.480***
SPU (mg g^{-1})	0.023^{NS}	-0.107^{NS}

***P<0.001

Frequency of colonization was significantly correlated with measured variables such as growth, biomass, dry matter, yield and P content whereas, in the case of SPU an insignificant negative relationship was observed. Intensity of colonization also followed the same pattern of relationships except for the SPU, where an insignificant negative relationship was exhibited. Harikumar and Potty (2002) in a field study with 257 genetic stocks of sweet potato for mycorrhizal colonization and its response on growth of the crop observed that fungal characters showed a positive relationship with underground plant characters (root and tuber) only while, the above ground portion had no direct bearing on the interaction. By contrast, in the present study all the parameters examined showed significant correlation with the fungal variables further confirming the importance of indigenous AM endophytes on growth and nutrition of sesame.

7. 1. Conclusion

Screening of field grown sesame accessions revealed that genotype dependent variation exists in this crop. In general, both frequency and intensity of colonization were low in sesame accessions. Nevertheless the crop is greatly dependent on indigenous AM colonization for growth and nutrition as a significant positive relationship between plant and fungal characters could be drawn out through correlation analysis. In the present study, however attempt has not been made to understand the species infecting the accessions and whether a single species or a consortium of species contributes the infection. Detailed studies in this direction may yield useful information on host-fungus compatibility.

CHAPTER 8

Effect of AM inoculation, P fertilization and Irrigation on Growth, Yield, Biochemical Composition and Nutrition of Sesame

8. 1. AM colonization

Inoculation with AM fungi significantly improved the mycorrhizal status of field grown sesame (Vijayalekshmi and Rao, 1988). In most cases, high concentration of soil P resulted in reductions in AM colonizations. (Johnson and Pfleger, 1992). In the present study P application even at the recommended dose did not cause any apparent reduction in root colonization by AM fungi (Table 8.1). This is in agreement with the finding of Khalil et al. (1992) that even under conditions of very high fertility, (60.3 mg kg^{-1}, Bray P) roots of soybean were extremely (89%) colonized by AM fungi. The difference in root colonization between rainfed and irrigated plants was not significant, though a slight increase in $\% F$ was observed in rainfed plants. This could be ascribed to the reason that an occasional irrigation schedule (at 15 days interval) might not have created a waterlogged or anoxic condition (Sundaresan et al., 1987). Numerous greenhouse and field experiments have shown conclusively that plants colonized by AM fungi are much more efficient in taking up soil P than non AM plants, particularly under drought condition (Nelsen and Safir, 1982). Improved P nutrition by AM fungi during water deficit has been postulated as a potential mechanism for enhancing host plant drought tolerance. Irrespective of treatments, the highest AM colonization was observed at late stages of growth. This is in agreement with the finding of Abbot and Robson (1991) that growth stage of the crop has a profound influence on root colonization by AM fungi.

Table 8. 1. Effect of AM inoculation, P fertilization and irrigation on root colonization (%F) by AM fungi in sesame

P level	M_0		M_1	
	I_0	I_1	I_0	I_1
D1				
0	10.00	10.00	20.00	15.00
1	10.00	13.33	23.33	16.66
2	20.00	11.67	20.00	18.33
3	10.00	16.66	13.33	23.33
4	3.33	20.00	10.00	23.33
D2				
0	21.67	10.00	31.67	21.66
1	16.33	21.67	27.00	22.67
2	25.00	21.67	31.67	33.33
3	15.00	23.33	23.66	30.33
4	11.67	26.67	27.27	31.00
D3				
0	14.52	12.73	50.00	41.67
1	29.66	24.33	60.33	27.00
2	26.27	24.50	37.00	23.67
3	25.17	23.33	31.33	39.33
4	36.67	32.00	37.00	44.00
Significance				
	M (<0.001)***	M×P (<0.001)***	M×P×I (<0.001)***	M×P×I×D (<0.001) ***
	P (0.003) ***	M×I (0.014)**	M×P×D (<0.001)***	
	I (0.546) NS	M×D (0.001) ***	M×I×D (0.485) NS	
	D (<0.001)***			

P values given in parenthesis
*Asterisk shows significance

8. 2. Growth and biomass production

Growth characters such as root length, shoot length, leaf number, biomass and dry matter production were significantly higher in AM inoculated treatments compared to uninoculated controls (Table 8. 2 – 7) (Chiramel et al., 2006).

Table 8. 2. Effect of AM inoculation, P fertilization and irrigation on root length (cm) by AM fungi in sesame

P level	M_0		M_1	
	I_0	I_1	I_0	I_1
D1				
0	03.25	03.73	02.67	03.67
1	03.50	02.00	01.83	03.33
2	04.33	02.33	05.17	03.67
3	04.00	04.67	04.00	04.00
4	03.00	04.00	04.00	05.43
D2				
0	07.00	05.33	08.00	05.00
1	06.97	06.00	05.33	04.00
2	05.90	04.66	05.67	05.00
3	05.13	05.33	07.17	03.00
4	08.33	04.33	04.70	05.33
D3				
0	07.17	04.67	05.67	04.67
1	06.00	06.33	06.00	03.50
2	04.33	03.33	13.00	11.67
3	04.00	04.33	07.00	05.00
4	04.33	03.67	08.50	08.33
Significance				
	M (0.001)***	M×P (<0.001)***	M×P×I (<0.001)***	M×P×I×D (<0.001)***
	P (<0.001)***	M×I (0.888) NS	M×P×D (<0.001)***	
	I (<0.001)***	M×D (<0.001)***	M×I×D (0.098) NS	
	D (<0.001)***			

P values given in parenthesis
*Asterisk shows significance

Table 8. 3. Effect of AM inoculation, P fertilization and irrigation on shoot length (cm) in sesame

P level	M_0		M_1	
	I_0	I_1	I_0	I_1
D1				
0	06.40	07.00	06.83	08.17
1	06.33	05.50	07.00	06.67
2	05.17	05.67	06.00	08.17
3	06.00	08.17	06.67	08.67
4	04.00	06.83	09.17	07.33
D2				
0	14.00	21.67	23.00	24.67
1	21.67	19.67	24.33	23.33
2	16.67	16.00	19.33	24.33
3	17.17	20.33	22.67	22.33
4	21.00	14.00	22.00	26.00
D3				
0	34.00	27.67	36.00	52.67
1	25.33	42.33	46.00	53.67
2	35.33	34.67	36.33	48.33
3	25.00	42.00	42.00	46.00
4	28.50	36.00	35.33	38.33
Significance				
	M (<0.001)***	M×P (<0.083) NS	M×P×I (<0.001)***	M×P×I×D (P<0.001)***
	P (0.001) ***	M×I (0.126) NS	M×P×D (<0.001)***	
	I (<0.001)***	M×D (0.001) ***	M×I×D (0.322) NS	
	D (<0.001)***			

P values given in parenthesis
*Asterisk shows significance

Table 8.4. Effect of AM inoculation, P fertilization and irrigation on leaf number in sesame.

P level	M_0		M_1	
	I_0	I_1	I_0	I_1
D1				
0	05.33	10.00	05.67	07.33
1	03.67	09.33	03.67	09.67
2	05.33	07.00	06.33	07.33
3	04.67	08.33	05.33	08.67
4	06.00	04.67	06.33	06.00
D2				
0	10.00	13.67	18.00	12.67
1	13.33	15.00	14.66	14.66
2	11.33	12.66	23.33	13.00
3	10.66	11.33	10.67	12.00
4	14.00	08.67	14.67	19.00
D3				
0	29.00	08.67	24.66	14.67
1	27.33	14.00	27.00	16.33
2	16.66	15.33	29.33	23.33
3	18.00	12.67	24.00	24.00
4	16.67	17.33	32.67	20.67
Significance				
	M (<0.001)***	M×P (<0.001)***	M×P×I (<0.004)**	M×P×I×D (<0.001) ***
	P (0.005) **	M×I (0.153) ^{NS}	M×P×D (<0.001)***	
	I (0.001)***	M×D (0.001) ***	M×I×D (0.228) ^{NS}	
	D (<0.001)***			

P values given in parenthesis
*Asterisk shows significance

Table 8. 5. Effect of AM inoculation, P fertilization and irrigation on leaf area (cm^2) in sesame

P level	M_0		M_1	
	I_0	I_1	I_0	I_1
D1				
0	12.86	39.79	44.43	65.00
1	15.55	71.00	19.53	71.33
2	36.33	38.20	42.20	44.70
3	26.08	28.20	27.73	45.17
4	54.80	42.20	60.93	52.20
D2				
0	39.73	49.33	152.33	111.00
1	88.86	42.33	92.93	177.33
2	72.46	27.50	144.33	70.33
3	30.53	38.66	39.13	94.67
4	81.93	54.20	87.80	119.00
D3				
0	229.33	40.30	425.00	75.67
1	317.33	27.56	311.67	128.67
2	102.00	56.60	188.33	146.00
3	155.33	126.07	165.33	243.00
4	138.00	137.23	171.67	152.86
Significance				
	M (<0.001)***	M×P (<0.001)***	M×P×I (<0.001)***	M×P×I×D (<0.001) ***
	P (0.001) ***	M×I (0.004)**	M×P×D (<0.001)***	
	I (0<0.001)***	M×D (0.001) ***	M×I×D (0.027)*	
	D (<0.001)***			

P values given in parenthesis
*Asterisk shows significance

Table 8.6. Effect of AM inoculation, P fertilization and irrigation on biomass production (g plant^{-1}) in sesame.

P level	M$_0$		M$_1$	
	I$_0$	I$_1$	I$_0$	I$_1$
D1				
0	01.24	01.54	04.79	01.81
1	01.70	01.38	04.04	02.58
2	01.21	01.27	01.82	01.57
3	01.92	01.93	03.47	02.35
4	01.35	01.48	01.95	03.13
D2				
0	01.87	00.35	02.11	01.63
1	01.13	00.54	01.69	02.28
2	01.25	00.34	02.50	01.15
3	00.80	00.48	02.22	01.22
4	01.71	01.08	01.67	01.76
D3				
0	01.86	01.18	03.93	03.42
1	01.56	01.26	04.72	03.84
2	01.76	01.44	03.20	04.33
3	01.38	02.23	02.31	02.49
4	01.89	01.32	02.65	01.75
Significance				
	M (<0.001)***	M×P (<0.001)***	M×P×I (0.004)**	M×P×I×D (<0.001)***
	P (0.001)***	M×I (0.119)^{NS}	M×P×D (<0.001)***	
	I (<0.001)***	M×D (<0.001)***	M×I×D (<0.001)***	
	D (<0.001)***			

P values given in parenthesis
*Asterisk shows significance

Table 8. 7. Effect of AM inoculation, P fertilization and irrigation on dry matter production (g plant^{-1}) in sesame

P level	M_0		M_1	
	I_0	I_1	I_0	I_1
D1				
0	00.22	00.27	00.47	00.32
1	00.28	00.27	00.47	00.39
2	00.28	00.25	00.23	00.28
3	00.34	00.34	00.52	00.35
4	00.26	00.29	00.31	00.36
D2				
0	00.39	00.11	00.40	00.42
1	00.30	00.13	00.39	00.48
2	00.28	00.05	00.43	00.24
3	00.18	00.10	00.36	00.30
4	00.36	00.22	00.25	00.29
D3				
0	01.11	00.23	02.07	00.50
1	00.83	00.67	03.16	02.48
2	01.05	00.95	02.19	02.72
3	00.97	01.61	01.44	01.74
4	01.29	00.63	01.24	01.20
Significance				
	M (<0.001)***	M×P (<0.001)***	M×P×I (<0.001)***	M×P×I×D (<0.005)**
	P (0.001)***	M×I (0.858)^{NS}	M×P×D (<0.001)***	
	I (<0.001)***	M×D (<0.001)***	M×I×D (0.163)^{NS}	
	D (<0.001)***			

P values given in parenthesis
*Asterisk shows significance

In most cases, a better response on growth characters was achieved at low to moderate level of P application. It is well known that under low levels of external P fertilizer, the fungal endophyte allows the development of more external hyphae (Schwab et al., 1983) for scavenging more water and nutrients (Clapperton and Reid, 1992) which in turn reflects on growth. Better growth response under optimal P fertilizer levels has been reported in crops such as citrus (Antunes and Cardoso, 1991) and canna (Jayakumari and Potty, 2009). A comparison of growth characters under rainfed and irrigated conditions revealed a general increment in growth characters under rainfed conditions. Enhanced growth and biomass production could be related to improved uptake of P and Ca especially under water stress condition (Al-Karaki and Clark, 1998). M×P×I and M×P×I interactions were significant for most of the growth characters measured. Further, the growth characters showed a general increment at advanced stages of crop growth.

8. 3. Yield and quality attributes

Mycorrhizal inoculation, P fertilization and irrigation significantly improved yield and quality attributes of sesame (Table 8. 8 – 13). However, the flower number did not show any significant response to AM inoculation, P fertilization or M×I and M×I×D interactions. However, M×P×I and M×P×I×D interactions were significant for flower number. Pod and seed number, 1000 seed weight and oil content were significantly high in AM inoculated plants. This could be attributed to higher growth stimulation by inoculation that proceeded to increased synthesis of food material by photosynthesis and their translocation to the reproductive sites, which maintained their favourable source-sink relationship (Singh et al., 2004).

Table 8.8. Flower number as affected by AM inoculation, P fertilization and irrigation in sesame

P level	M_0		M_1	
	I_0	I_1	I_0	I_1
D1				
0	00.00	00.00	00.00	00.00
1	00.00	00.00	00.00	00.00
2	00.00	00.00	00.00	00.00
3	00.00	00.00	00.00	00.00
4	00.00	00.00	00.00	00.00
D2				
0	05.00	03.00	03.33	01.67
1	06.67	03.00	02.33	01.00
2	02.00	01.67	08.67	02.00
3	02.00	01.33	03.33	01.67
4	06.00	02.00	02.67	02.67
D3				
0	00.00	00.00	01.33	00.00
1	00.00	00.00	00.67	00.00
2	00.00	00.00	00.00	00.00
3	00.00	00.00	01.00	00.00
4	00.00	00.00	00.67	00.00
Significance				
	M (0.930) ^{NS}	M×P (<0.001)***	M×P×I (<0.001)***	M×P×I×D (<0.001)***
	P (0.175) ^{NS}	M×I (0.253) ^{NS}	M×P×D (<0.001)***	
	I (<0.001)***	M×D (<0.001)***	M×I×D (0.450) ^{NS}	
	D (<0.001)***			

P values given in parenthesis
*Asterisk shows significance

Table 8. 9. Pod number as affected by AM inoculation, P fertilization and irrigation in sesame

P level	M_0		M_1	
	I_0	I_1	I_0	I_1
D1				
0	00.00	00.00	00.00	00.00
1	00.00	00.00	00.00	00.00
2	00.00	00.00	00.00	00.00
3	00.00	00.00	00.00	00.00
4	00.00	00.00	00.00	01.33
D2				
0	01.00	01.00	00.33	02.00
1	00.67	01.00	01.33	02.33
2	01.33	01.67	02.33	02.33
3	00.00	01.33	02.00	01.67
4	01.33	00.00	01.33	00.00
D3				
0	05.67	03.33	06.33	04.00
1	05.33	04.33	05.66	05.00
2	04.67	02.67	06.00	05.33
3	04.33	04.67	04.67	05.33
4	03.33	03.33	05.33	04.33
Significance				
	M (<0.001)***	M×P (<0.313) ^{NS}	M×P×I (0.592) ^{NS}	M×P×I×D (0.100) ^{NS}
	P (0.024) *	M×I (0.384) ^{NS}	M×P×D (<0.047)*	
	I (0.52)*	M×D (0.002) **	M×I×D (0.920) ^{NS}	
	D (<0.001)***			

P values given in parenthesis
*Asterisk shows significance

Table 8. 10. Seed Number (plant^{-1}) as affected by AM inoculation P fertilization and irrigation in sesame

P level	M_0		M_1	
	I_0	I_1	I_0	I_1
D1				
0	00.00	00.00	00.00	00.00
1	00.00	00.00	00.00	00.00
2	00.00	00.00	00.00	00.00
3	00.00	00.00	00.00	00.00
4	00.00	00.00	00.00	00.00
D2				
0	00.00	00.00	00.00	00.00
1	00.00	00.00	00.00	00.00
2	00.00	00.00	00.00	00.00
3	00.00	00.00	00.00	00.00
4	00.00	00.00	00.00	00.00
D3				
0	63.33	49.00	93.33	58.33
1	72.33	90.00	176.33	124.00
2	120.67	60.00	155.00	176.77
3	28.33	68.67	99.33	82.00
4	52.00	72.00	73.67	70.33
Significance				
	M (<0.001)***	M×P (<0.001)***	M×P×I (<0.001)***	M×P×I×D (<0.001)***
	P (<0.001)***	M×I (0.026)*	M×P×D (<0.001)***	
	I (0.038)*	M×D (<0.001)***	M×I×D (0.008)**	
	D (<0.001)***			

P values given in parenthesis
*Asterisk shows significance

Table 8. 11. Seed weight (g plant^{-1}) as affected by AM inoculation, P fertilization and irrigation in sesame.

P level	M_0		M_1	
	I_0	I_1	I_0	I_1
D1				
0	00.00	00.00	00.00	00.00
1	00.00	00.00	00.00	00.00
2	00.00	00.00	00.00	00.00
3	00.00	00.00	00.00	00.00
4	00.00	00.00	00.00	00.00
D2				
0	00.00	00.00	00.00	00.00
1	00.00	00.00	00.00	00.00
2	00.00	00.00	00.00	00.00
3	00.00	00.00	00.00	00.00
4	00.00	00.00	00.00	00.00
D3				
0	00.24	00.15	00.35	00.45
1	00.26	00.29	00.63	00.37
2	00.36	00.23	00.46	00.53
3	00.27	00.31	00.39	00.23
4	00.16	00.22	00.22	00.26
Significance				
	M (<0.001)***	M×P (<0.001)***	M×P×I (<0.001)***	M×P×I×D (<0.001) ***
	P (<0.001) ***	M×I (0.429) ^{NS}	M×P×D (<0.001)***	
	I (0.083) ^{NS}	M×D (<0.001) ***	M×I×D (0.010)**	
	D (<0.001)***			

P values given in parenthesis
*Asterisk shows significance

Table 8. 12. 1000 seed weight as affected by AM inoculation P fertilization and irrigation in sesame

P level	M_0		M_1	
	I_0	I_1	I_0	I_1
D1	00.00	00.00	00.00	00.00
0	00.00	00.00	00.00	00.00
1	00.00	00.00	00.00	00.00
2	00.00	00.00	00.00	00.00
3	00.00	00.00	00.00	00.00
4				
D2				
0	00.00	00.00	00.00	00.00
1	00.00	00.00	00.00	00.00
2	00.00	00.00	00.00	00.00
3	00.00	00.00	00.00	00.00
4	00.00	00.00	00.00	00.00
D3				
0	03.81	03.06	03.61	07.67
1	03.57	03.30	03.58	02.98
2	03.84	04.18	02.99	03.00
3	08.35	04.72	03.96	02.85
4	03.01	03.06	02.98	03.67
Significance				
	M (0.067) ^{NS}	M×P (<0.001)***	M×P×I (<0.001)***	M×P×I×D (<0.001) ***
	P (<0.001) ***	M×I (<0.01)***	M×P×D (<0.001)***	
	I (0.532) ^{NS}	M×D (0.036) *	M×I×D (<0.001)***	
	D (<0.001)***			

P values given in parenthesis
*Asterisk shows significance

Table 8. 13. Oil content (%) as affected by AM inoculation, P fertilization and irrigation in sesame

P level	M_0		M_1	
	I_0	I_1	I_0	I_1
D1	00.00	00.00	00.00	00.00
0	00.00	00.00	00.00	00.00
1	00.00	00.00	00.00	00.00
2	00.00	00.00	00.00	00.00
3	00.00	00.00	00.00	00.00
4				
D2				
0	00.00	00.00	00.00	00.00
1	00.00	00.00	00.00	00.00
2	00.00	00.00	00.00	00.00
3	00.00	00.00	00.00	00.00
4	00.00	00.00	00.00	00.00
D3				
0	27.16	26.67	46.63	27.37
1	30.60	35.10	57.23	33.27
2	29.93	30.83	29.13	28.60
3	30.10	44.33	61.67	38.20
4	40.27	28.40	39.83	46.20
Significance				
	M (<0.001)***	M×P (<0.025)*	M×P×I (<0.001)***	M×P×I×D (<0.001) ***
	P (<0.001) ***	M×I (0.001)**	M×P×D (<0.006)**	
	I (0.001)***	M×D (0.001) ***	M×I×D (<0.001)***	
	D (<0.001)***			

P values given in parenthesis
*Asterisk shows significance

Among the different levels of P fertilization, the highest yield response was observed under moderate to slightly higher (11.25 kgh^{-1}) levels of P application, suggesting the increased demand of P during grain filling (Eid et al., 1951; Plenchette et al., 1983). In most cases yield maximization was achieved in rainfed plants, possibly due to an enhanced nutrient uptake in water stressed condition (Sieverding and Syverston, 1983). M×P×I and M×P×I×D interactions were significant for yield and quality attributes showing that these interactions could improve the yield and quality attributes of sesame.

8. 4. Biometric parameters

In general, inoculation with AM fungi significantly improved the biometric parameters in sesame (Table 8. 14 – 18). However the response of P levels and irrigation on biometric parameters were case dependent. For example, LAI showed maximum values at low level of P while in the case of NAR addition of P did not give any significant effect. SLW increased at both lower and higher levels of P application. Other parameters (HI and OI) showed a better response at moderate level of added P. The inconsistency in the biometric parameters observed in the present study is likely to be due to the differences in the uptake of nutrients and water by the hyphal system (Estrada-Luna and Davis, 2003).

Table 8. 14. Leaf area index as influenced by AM inoculation, P fertilization and irrigation in sesame.

P level	M_0		M_1	
	I_0	I_1	I_0	I_1
D1				
0	0.85	2.65	02.93	04.27
1	1.03	4.73	01.25	04.73
2	2.42	2.53	02.81	02.90
3	1.74	1.83	01.82	03.00
4	3.65	2.77	04.02	03.43
D2				
0	02.65	03.27	10.15	07.53
1	05.92	02.80	06.17	11.77
2	04.81	01.83	10.03	04.63
3	02.03	02.53	02.61	06.30
4	05.46	03.60	05.80	07.93
D3				
0	09.62	02.67	28.30	05.00
1	21.13	01.80	20.73	08.53
2	06.80	03.73	12.53	09.70
3	10.33	08.37	10.97	16.17
4	09.17	09.13	11.43	10.13
Significance				
	M (<0.001)***	M×P (<0.001)***	M×P×I (<0.001)***	M×P×I×D (<0.001) ***
	P (<0.001) ***	M×I (0.239) NS	M×P×D (<0.001)***	
	I (<0.001)***	M×D (0.001) ***	M×I×D (0.043)*	
	D (<0.001)***			

P values given in parenthesis
*Asterisk shows significance

Table 8. 15. Specific leaf weight (mg cm^{-2}) as influenced by AM inoculation, P fertilization and irrigation in sesame.

P level	M_0		M_1	
	I_0	I_1	I_0	I_1
D1				
0	00.45	00.15	00.27	00.43
1	00.63	00.11	00.47	00.30
2	00.14	00.14	00.17	00.24
3	00.29	00.16	00.27	00.21
4	00.18	00.20	00.57	00.35
D2				
0	01.06	00.22	00.52	00.25
1	00.31	00.19	00.75	00.29
2	00.81	00.36	00.41	00.37
3	00.63	00.26	00.54	00.26
4	00.18	00.66	00.52	00.96
D3				
0	00.58	00.67	00.54	00.76
1	00.39	00.84	00.42	00.63
2	00.48	00.64	00.53	01.18
3	00.73	00.21	00.35	00.56
4	00.73	00.47	00.90	00.43
Significance				
	M (0.015)*	M×P (0.001)***	M×P×I (<0.001)***	M×P×I×D (<0.001)***
	P (0.001)***	M×I (<0.001)***	M×P×D (<0.001)***	
	I (0.<0.001)***	M×D (0.509) NS	M×I×D (0.898) NS	
	D (<0.001)***			

P values given in parenthesis
*Asterisk shows significance

Table 8. 16. Net assimilation rate (g dm^{-2} day^{-1}) as influenced by AM inoculation, P fertilization and irrigation in sesame

P level	M_0		M_1	
	I_0	I_1	I_0	I_1
D1			00.64	00.44
0	00.90	00.43	01.16	00.97
1	00.97	00.37	00.53	00.60
2	00.37	00.33	01.18	00.87
3	00.89	00.42	00.99	00.49
4	00.32	00.45		
D2				
0	-02.43	00.01	00.08	00.05
1	-00.22	00.03	00.02	00.02
2	00.03	-00.05	00.14	00.14
3	00.13	00.02	00.05	00.14
4	00.05	00.03	-00.01	00.02
D3				
0	00.27	00.18	00.07	00.50
1	00.67	00.21	00.11	00.30
2	00.16	00.17	01.00	00.42
3	00.11	00.21	00.12	00.24
4	00.11	00.19	00.19	00.14
Significance				
	M (<0.001)***	M×P (0.659)^{NS}	M×P×I (0.485)^{NS}	M×P×I×D (0.082)^{NS}
	P (0.140)^{NS}	M×I (0.506)^{NS}	M×P×D (0.051)*	
	I (0.898)^{NS}	M×D (0.490)^{NS}	M×I×D (0.254)^{NS}	
	D (<0.001)***			

P values given in parenthesis
*Asterisk shows significance

Table 8. 17. Harvest index (%) as influenced by AM inoculation, P fertilization and irrigation in sesame.

P level	M_0		M_1	
	I_0	I_1	I_0	I_1
D1				
0	00.00	00.00	00.00	00.00
1	00.00	00.00	00.00	00.00
2	00.00	00.00	00.00	00.00
3	00.00	00.00	00.00	00.00
4	00.00	00.00	00.00	00.00
D2				
0	00.00	00.00	00.00	00.00
1	00.00	00.00	00.00	00.00
2	00.00	00.00	00.00	00.00
3	00.00	00.00	00.00	00.00
4	00.00	00.00	00.00	00.00
D3				
0	00.12	00.11	00.08	00.12
1	00.15	00.19	00.12	00.09
2	00.17	00.13	00.13	00.11
3	00.16	00.12	00.14	00.09
4	00.08	00.14	00.08	00.13
Significance				
	M (<0.001)***	M×P (<0.068) ^{NS}	M×P×I (0.121) ^{NS}	M×P×I×D (0.072) ^{NS}
	P (0.004) **	M×I (0.447) ^{NS}	M×P×D (0.028)*	
	I (0.800) ^{NS}	M×D (<0.001) ***	M×I×D (0.560) ^{NS}	
	D (<0.001)***			

P values given in parenthesis
*Asterisk shows significance

Table 8. 18. Oil index (g oil 1000^{-1}) as influenced by AM inoculation, P fertilization and irrigation in sesame

P level	M_0		M_1	
	I_0	I_1	I_0	I_1
D1				
0	00.00	00.00	00.00	00.00
1	00.00	00.00	00.00	00.00
2	00.00	00.00	00.00	00.00
3	00.00	00.00	00.00	00.00
4	00.00	00.00	00.00	00.00
D2				
0	00.00	00.00	00.00	00.00
1	00.00	00.00	00.00	00.00
2	00.00	00.00	00.00	00.00
3	00.00	00.00	00.00	00.00
4	00.00	00.00	00.00	00.00
D3				
0	01.17	00.81	01.70	02.11
1	00.78	01.14	01.99	00.99
2	01.15	01.27	00.96	00.85
3	02.51	02.08	02.43	01.09
4	01.21	00.87	01.89	01.71
Significance				
	M (0.027)*	M×P (<0.001)***	M×P×I (<0.001)***	M×P×I×D (<0.001) ***
	P (<0.001) ***	M×I (0.346) NS	M×P×D (<0.001)***	
	I (0.021)*	M×D (<0.001) ***	M×I×D (0.411) NS	
	D (<0.001)***			

P values given in parenthesis
*Asterisk shows significance

Understanding the mechanism associated with this pattern will require further work. The interactions (M×P, M×P×I, M×P×I×D) effects were significant only in the case of LAI, SLW and HI. It is well known that AM fungi improve P availability to host plants and alter their morphology, physiology and competitive ability. Increase in biometric parameters consequent to AM inoculation has been corroborated by several workers (Jakobsen, 1987; Bray et al., 2003; Morone-Fortunato and Avato, 2008).

8. 5. Biochemical characteristics

Biochemical characteristics of the plants varied significantly due to AM colonization, P fertilization and irrigation (Table 8.19 – 22). Tissue CHO content was significantly higher in AM inoculated plants (Khalafallah and Abo-Ghalia, 2008). CHO content was more under low levels of P fertilization indicating that high soil P disrupts the CHO formation in tissue (Same et al., 1983). Irrigation improved the CHO content in plants. Sorial (2001) observed that AM inoculation recorded highly significant induction of total sugars under water stress condition. Interaction of AM fungi, P fertilization and irrigation had a significant effect in enhancing this parameter during different stages of growth.

AM inoculation also assisted the host plant to maintain higher protein concentration in tissue as compared with uninoculated control (Subramanian and Charest, 1999). Protein content in tissue varied significantly under varied P levels. However, extra input of P did not make any added benefit on the quantitative increase of protein over control. Irrigation slightly improved the protein content in tissue while M×I interaction had no significant effect on tissue protein content.

Amino acid content in plant tissue was significantly improved in AM inoculated plants (Valentine et al., 2006). Increased amino acid content was observed in the treatments received lower to medium level of P application showing that availability of P beyond certain level could limit amino acid synthesis in the crop. A similar increase in tissue amino acid content at ½ the recommended dose of P (KH_2PO_4) was observed in *Phaseolus vulgaris* by Neeraj and Singh (2008).

Table 8. 19. Effect of AM inoculation, P fertilization and irrigation on total carbohydrate content (mg g^{-1}) in sesame.

P level	M_0		M_1	
	I_0	I_1	I_0	I_1
D1				
0	85.37	99.30	259.33	135.20
1	95.00	153.23	172.67	192.33
2	104.00	72.70	223.43	88.70
3	124.27	98.73	95.47	199.37
4	58.53	125.13	146.50	70.97
D2				
0	49.87	121.93	229.40	204.00
1	193.97	74.70	197.10	188.67
2	146.40	98.00	61.60	304.67
3	202.33	78.76	43.80	539.23
4	91.90	76.20	75.90	248.13
D3				
0	187.87	140.27	250.00	125.73
1	237.33	119.87	129.20	184.33
2	199.80	200.73	188.87	130.13
3	125.47	227.33	60.40	121.07
4	227.00	228.67	27.00	33.07
Significance				
	M (<0.001)***	M×P (<0.001)***	M×P×I (<0.001)***	M×P×I×D (<0.001) ***
	P (<0.001) ***	M×I (<0.001)***	M×P×D (<0.001)***	
	I (<0.001)***	M×D (<0.001) ***	M×I×D (<0.001)***	
	D (<0.001)***			

P values given in parenthesis
*Asterisk shows significance

Table 8. 20. Effect of AM inoculation, P fertilization and irrigation on protein content (mg g^{-1}) in sesame.

P level	M_0		M_1	
	I_0	I_1	I_0	I_1
D1				
0	13.63	02.30	27.43	22.33
1	03.67	12.30	30.50	14.33
2	06.70	14.10	6.57	14.30
3	03.10	16.33	5.87	20.97
4	03.60	06.30	38.00	12.80
D2				
0	32.63	17.03	39.50	67.86
1	31.27	10.27	33.13	44.93
2	26.33	34.53	26.63	42.30
3	28.47	37.37	23.57	29.90
4	28.56	25.87	26.93	29.53
D3				
0	01.90	02.87	09.43	08.90
1	13.13	11.97	13.73	09.90
2	12.80	16.07	13.83	12.17
3	08.17	09.37	09.00	09.17
4	00.23	11.37	10.87	02.07
Significance				
	M (<0.001)***	M×P (<0.001)***	M×P×I (<0.001)***	M×P×I×D (<0.001)***
	P (<0.001)***	M×I (0.342) NS	M×P×D (<0.001)***	
	I (0.003)***	M×D (<0.001)***	M×I×D (<0.001)***	
	D (<0.001)***			

P values given in parenthesis
*Asterisk shows significance

Table 8. 21. Effect of AM inoculation, P fertilization and irrigation on amino acid content in (mg g^{-1}) in sesame.

P level	M$_0$		M$_1$	
	I$_0$	I$_1$	I$_0$	I$_1$
D1				
0	13.90	46.60	21.70	95.57
1	48.80	80.96	14.27	35.44
2	37.13	33.63	83.30	114.33
3	45.99	42.03	35.30	41.03
4	12.73	2.33	90.93	69.33
D2				
0	07.86	38.31	9.49	135.93
1	27.33	49.39	28.00	55.00
2	50.10	66.61	68.93	44.12
3	27.60	13.79	27.87	95.93
4	43.70	07.97	48.90	129.43
D3				
0	20.73	06.47	20.76	03.87
1	00.47	17.87	17.26	12.93
2	10.20	39.07	12.40	11.27
3	20.00	11.53	16.93	13.73
4	22.40	22.47	3.00	23.07
Significance				
	M (<0.001)***	M×P (<0.001)***	M×P×I (<0.001)***	M×P×I×D (<0.001) ***
	P (<0.001) ***	M×I (<0.001)***	M×P×D (<0.001)***	
	I (<0.001)***	M×D (<0.001) ***	M×I×D (<0.001)***	
	D (<0.001)***			

P values given in parenthesis
*Asterisk shows significance

Table 8. 22. Effect of AM inoculation P fertilization and irrigation on proline content (mg g^{-1}) in sesame

P level	M$_0$		M$_1$	
	I$_0$	I$_1$	I$_0$	I$_1$
D1				
0	28.80	24.50	28.43	15.73
1	11.27	18.20	40.00	16.97
2	21.80	11.83	35.77	11.57
3	09.17	18.13	25.73	16.73
4	19.23	34.87	37.50	20.33
D2				
0	32.17	20.67	26.70	19.73
1	13.87	17.89	08.47	29.40
2	19.67	17.20	09.43	34.77
3	57.87	17.63	06.90	17.83
4	12.53	23.93	13.30	13.43
D3				
0	02.40	01.33	02.13	01.13
1	02.40	02.97	01.67	01.33
2	03.80	01.40	01.60	02.47
3	01.33	01.87	01.20	02.87
4	02.93	01.93	02.00	02.27
Significance				
	M (0.633)^{NS}	M×P (<0.001)***	M×P×I (<0.001)***	M×P×I×D (<0.001)***
	P (<0.001)***	M×I (0.469)^{NS}	M×P×D (<0.001)***	
	I (<0.001)***	M×D (<0.001)***	M×I×D (<0.001)***	
	D (<0.001)***			

P values given in parenthesis
*Asterisk shows significance

Irrigation significantly improved amino acid content in tissue (Sorial, 2001). In general, high amino acid content was recorded during the early stages of plant growth. All the interactions involving mycorrhiza, P fertilization and irrigation were found to be significant for amino acid content in plants.

Drought stress is known to result in high proline content in AM plants. By contrast, in the present study the AM inoculated plants had a low proline quantity indicating that the plants were not prone to drought stress. Zézé et al. (2001) observed a low level of proline in the tissue of *Trifolium alexandrium* grown under water stress condition.

8. 6. Tissue P content

Phosphorus content of the tissue increased up to 41.8% in AM inoculated plants as compared to uninoculated control (Table 8.23). Chiramel et al. (2006) in a study of the response of *Andrographis paniculata* to different AM fungi observed an increase in shoot and root P concentration of 10.6 and 4.4% respectively in *G. leptoticum* inoculated plants and 12.8 and 3.6% in *G. intraradices* inoculated plants compared with uninoculated plants. It is noteworthy that the plants which received a relatively low level of fertilizer P had the highest P indicating that the tissue P content need not necessarily be related to the amount of P applied to soil. This contrasts the finding of Van and Tran (1990) that a linear relationship exists between soil available P and tissue P. Comparison of rainfed and irrigated plants revealed a high rate of P content in rainfed plants which is in agreement with the finding of Subramanian and Charest (1999) that AM colonization conferred a higher P status under drought condition. This could be due to be a better harvest of P from soil by hyphal extensions. Sundaresan et al. (1987) suggested that high soil moisture is not conducive for AM colonization due to water logging and oxygen deficiency. The M×I interaction was not significant for tissue P content. Tissue P content declined with plant age.

Table 8. 23. Effect of AM inoculation, P fertilization and irrigation on tissue P (mg g^{-1}) content in sesame

P level	M_0		M_1	
	I_0	I_1	I_0	I_1
D1				
0	01.87	01.55	17.37	03.00
1	01.47	01.90	16.00	08.07
2	10.83	07.90	08.43	06.33
3	07.80	02.35	13.90	21.93
4	12.40	07.47	12.40	03.50
D2				
0	05.87	04.40	05.10	04.53
1	07.00	01.90	08.17	02.79
2	04.40	03.67	01.53	11.67
3	12.63	04.70	24.33	13.40
4	04.67	03.90	07.13	06.20
D3				
0	01.87	01.87	05.20	01.40
1	01.87	01.40	01.87	01.50
2	03.27	01.87	02.37	02.46
3	02.23	02.33	02.33	01.98
4	02.33	01.87	03.27	04.47
Significance				
	M (<0.001)***	M×P (<0.001)***	M×P×I (<0.001)***	M×P×I×D (<0.001)***
	P (<0.001)***	M×I (0.444)^{NS}	M×P×D (<0.001)***	
	I (<0.001)***	M×D (<0.001)***	M×I×D (<0.001)***	
	D (<0.001)***			

P values given in parenthesis
*Asterisk shows significance

8. 7. Conclusion

The present study envisaged the beneficial effect of AM inoculation, P fertilization and irrigation or growth, yield biochemical composition and P nutrition in sesame. AM inoculation improved most of the measured parameters at the minimal level of P input and irrigation suggesting that the interaction involving AM fungi P fertilizer and irrigation could help the crop in achieving better growth, yield response and P nutrition.

CHAPTER 9

Individual and Interactive Effects of AM fungus *G. dimorphicum*, *Azospirillum* and Fertilizer in the Rhizosphere of Sesame and its Effect on Growth, Yield and Nutrition

9. 1. AM colonisation

Inoculation with AM fungus *G. dimorphicum* significantly improved the mycorrhizal status of sesame plants (Table 9. 1). It has been observed in several instances that introduced AM fungi colonize plant roots to a greater degree than the native ones (Schreiner, 2007). Therefore the increase in colonization level observed in the present study is not surprising. Maximum colonization (% *F*) was observed when the soil was inoculated with AM fungi at 50 percent recommended dose of NP fertilizer. Further, the plants inoculated with *Azospirillum* had more colonization as compared to inoculated ones. This contracts the finding of Russo et al. (2005) in maize and wheat that the wild type or genetically modified *Azospirillum* phytostimulators does not alter mycorrhization. More root colonization was observed during the late stages of plant growth. The study of the interactive effects revealed that M×A combinations had no significant effect on root colonization by AM fungi (Russo et al., 2005). However, the presence of NP fertilizers had a stimulating effect in augmenting the root colonization. The interactions involving AM fungi, *Azospirillum* and NP fertilizer was significant with plant growth stages (M×A×F×D) indicating the stimulatory effect of the combination on root colonization.

Table. 9. 1. Effect of AM fungi, *Azospirillum* and fertilizer on mycorrhizal colonization *(% F)* in sesame.

AM level	*Azo.* level	Fert. level	% F	
			35 DAS	**70 DAS**
M_0	A_0	0	21.67	18.33
		1	23.33	25.00
		2	21.66	28.33
		3	11.67	22.33
		4	10.67	24.33
	A_1	0	30.00	28.66
		1	30.00	20.67
		2	23.33	26.67
		3	20.00	16.00
		4	15.00	34.33
M_1	A_0	0	45.00	32.00
		1	50.00	31.33
		2	46.67	72.33
		3	41.67	37.67
		4	38.33	32.66
	A_1	0	48.33	41.00
		1	53.33	61.00
		2	43.33	69.33
		3	43.33	61.67
		4	35.00	45.00
Significance				
M			<0.001***	
F			<0.001***	
A			<0.001***	
D			0.001***	
M × F			<0.001***	
M × A			0.118^{NS}	
M × D			0.942^{NS}	
M × F × A			0.018*	
M × F × D			<0.001***	
M × A × D			<0.001***	
M × F × A × D			0.028*	

9. 2. Growth characteristics

Growth characteristics such as plant height, leaf number and leaf area were significantly improved due to AM inoculation (Table 9. 2).

Table. 9. 2. Growth characteristics as influenced by inoculation of AM fungi and *Azospirllum* under different fertilizer level in sesame.

AM level	Azo. level	Fert. level	Plant height (cm)		Leaf no.		Leaf area	
			35 DAS	70 DAS	35 DAS	70 DAS	35 DAS	70 DAS
M_0	A_0	0	15.67	19.33	8.67	14.00	20.19	89.43
		1	18.00	27.00	7.67	11.67	15.68	70.46
		2	16.67	22.67	8.67	17.33	16.66	148.83
		3	17.33	24.67	8.67	20.00	15.31	244.27
		4	14.00	21.00	6.67	13.33	12.78	91.07
	A_1	0	19.33	22.00	8.67	18.67	28.91	162.79
		1	20.33	28.67	8.00	15.00	30.91	111.78
		2	22.33	30.67	10.00	16.00	53.57	139.06
		3	17.67	24.67	7.00	15.33	20.08	89.17
		4	12.33	17.00	6.00	9.67	11.65	63.65
M_1	A_0	0	22.00	29.67	10.00	16.00	38.78	118.42
		1	22.00	29.00	8.33	14.33	23.80	136.50
		2	21.33	27.33	7.33	18.33	27.77	250.37
		3	21.33	27.33	7.33	13.67	23.33	133.57
		4	19.00	26.00	8.00	13.67	17.32	92.26
	A_1	0	19.33	32.67	8.00	12.67	30.76	99.35
		1	22.33	33.33	9.00	17.67	40.62	179.23
		2	22.33	26.33	8.67	19.00	28.23	219.12
		3	24.33	32.33	9.33	14.33	40.65	124.62
		4	18.67	26.33	9.33	14.37	31.80	94.06
Significance								
M			<0.000***		0.202^{NS}		<0.001***	
F			<0.000***		<0.001***		<0.001***	
A			0.001***		0.644^{NS}		0.856^{NS}	
D			<0.001***		<0.001***		<0.001***	
M × F			<0.001***		0.001***		<0.001***	
M × A			0.522^{NS}		0.240^{NS}		0.588^{NS}	
M × D			0.125^{NS}		0.719^{NS}		0.031*	
M × F × A			0.001***		<0.001***		<0.001***	
M × F × D			0.014**		0.003**		<0.001***	
M × A × D			0.165^{NS}		0.959^{NS}		0.240^{NS}	
M × F × A × D			0.357^{NS}		0.228^{NS}		0.001***	

Maximum effect of AM fungus on these parameters was observed when soil was inoculated with the fungus at moderate to higher levels of N and P suggesting the high demand of N and P during crop growth. The influence of AM fungus on increased plant growth is perhaps due to increased uptake of these nutrients, which might have caused cell elongation and multiplication (Black, 1965). Inoculation of *Azospirillum* alone did not make any enhancement of growth characteristics except the plant height. M×F×A interactions were significant for all the growth characteristics examined. However, the significance of the interaction with growth stages was observed only in the case of leaf area. Inoculation of plants with both *Azospirillum* and AM fungi have been shown to promote better plant growth (Kumutha et al., 1993) particularly in presence of P supporting the conclusion that P is required for N_2 fixation. The P made available by the AM fungi might have contributed to N_2 fixation in the presence of *Azospirillum*. Positive effects of combined inoculation with *Azospirillum* and different strains of AM fungi have been reported for barley (Negi et al., 1990) and sweet potato (Kandasamy et al., 1988).

9. 3. Yield and quality attributes

The data on the yield and quality attributes of sesame revealed a significant improvement consequent to AM inoculation (Table 9. 3 – 4) (Potty and Harikumar, 1995; Gupta et al., 2002).

Table. 9. 3. Effect of AM fungi, *Azospirillum* and fertilizer on the yield of sesame.

AM	Azo. level	Fert. level	Pod no.		Seed no.		Seed wt (g)		1000 seed wt.	
			35 DAS	70 DAS	35 DAS	70 DAS	35 DAS	70 DAS	35 DAS	70 DAS
M_0	A_0	0	0.00	1.00	0.00	10.00	0.00	0.04	0.00	1.00
		1	0.00	2.00	0.00	10.67	0.00	0.03	0.00	0.71
		2	0.00	1.67	0.00	14.00	0.00	0.05	0.00	0.81
		3	0.00	1.67	0.00	20.33	0.00	0.07	0.00	0.46
		4	0.00	3.00	0.00	48.00	0.00	0.23	0.00	0.83
	A_1	0	0.00	1.33	0.00	13.33	0.00	0.06	0.00	0.58
		1	0.00	1.67	0.00	14.00	0.00	0.03	0.00	0.37
		2	0.00	1.67	0.00	14.00	0.00	0.05	0.00	0.58
		3	0.00	2.67	0.00	27.33	0.00	0.10	0.00	0.65
		4	0.00	3.67	0.00	92.67	0.00	0.31	0.00	1.24
M_1	A_0	0	0.00	2.33	0.00	19.33	0.00	0.07	0.00	0.82
		1	0.00	3.33	0.00	28.67	0.00	0.10	0.00	0.56
		2	0.00	1.67	0.00	25.00	0.00	0.06	0.00	0.65
		3	0.00	3.33	0.00	58.33	0.00	0.21	0.00	1.79
		4	0.00	4.33	0.00	67.67	0.00	0.32	0.00	0.67
	A_1	0	0.00	2.67	0.00	31.67	0.00	0.09	0.00	1.11
		1	0.00	3.33	0.00	36.67	0.00	0.10	0.00	0.55
		2	0.00	2.67	0.00	38.33	0.00	0.09	0.00	0.54
		3	0.00	3.33	0.00	100.00	0.00	0.24	0.00	1.94
		4	0.00	6.67	0.00	125.37	0.00	0.30	0.00	1.67
Significance										
M			<0.001***		<0.001***		0.001***		<0.002**	
F			<0.001***		<0.001***		<0.001***		<0.001***	
A			0.015**		<0.001***		0.310^{NS}		0.328^{NS}	
D			<0.001***		<0.001***		<0.001***		<0.001***	
M × F			0.186^{NS}		0.015**		0.203^{NS}		<0.001***	
M × A			0.352^{NS}		0.072^{NS}		0.796^{NS}		0.071^{NS}	
M × D			<0.001***		<0.001***		0.001***		0.002**	
M × F × A			0.352^{NS}		0.811^{NS}		0.853^{NS}		0.703^{NS}	
M × F × D			$0,186^{NS}$		0.015**		0.203^{NS}		<0.001***	
M × A × D			0.352^{NS}		0.072^{NS}		0.796^{NS}		0.071^{NS}	
M × F × A × D			0.352^{NS}		0.811^{NS}		0.853^{NS}		0.703^{NS}	

Table. 9. 4. Quality attributes as influenced by inoculation of AM fungi and *Azospirillum* under different fertilizer level in sesame.

AM level	*Azo.* level	Fert. level	Oil content (%)		Oil index	
			35 DAS	70 DAS	35 DAS	70 DAS
M_0	A_0	0	0.00	32.00	0.00	1.17
		1	0.00	20.90	0.00	0.71
		2	0.00	32.67	0.00	0.81
		3	0.00	33.57	0.00	0.46
		4	0.00	31.84	0.00	0.83
	A_1	0	0.00	31.78	0.00	0.58
		1	0.00	12.70	0.00	0.37
		2	0.00	25.26	0.00	0.58
		3	0.00	24.67	0.00	0.65
		4	0.00	41.57	0.00	1.24
M_1	A_0	0	0.00	34.50	0.00	0.82
		1	0.00	36.00	0.00	0.56
		2	0.00	24.00	0.00	0.65
		3	0.00	62.00	0.00	1.79
		4	0.00	22.67	0.00	0.67
	A_1	0	0.00	38.67	0.00	1.11
		1	0.00	32.67	0.00	0.55
		2	0.00	21.00	0.00	0.54
		3	0.00	65.33	0.00	1.94
		4	0.00	57.67	0.00	1.65
Significance						
M			<0.001***		<0.004**	
F			<0.001***		<0.001***	
A			0.342^{NS}		0.454^{NS}	
D			<0.001***		<0.001***	
M × F			<0.001***		<0.001***	
M × A			0.023*		0.051*	
M × D			<0.001***		<0.004**	
M × F × A			0.507^{NS}		0.600^{NS}	
M × F × D			<0.001***		<0.001***	
M × A × D			0.023*		0.59^{NS}	
M × F × A × D			0.843^{NS}		0.600^{NS}	

In general, the yield and quality attributes were enhanced at high level (75% to full recommended dose) of added N and P possibly due to a better availability of nutrients in the rhizosphere which enabled an efficient translocation to the host by the mycorrhizal system. High requirement of N and P during the growth stages of plants has already been reported (Novoa and Loomis, 1981; Römer and Schilling, 1986). In the present study inoculation of *Azospirillum* alone did not have any beneficial effects on growth characteristics except pod number and seed number. This is in congruence with the finding of Fulchieri and Frioni (1994) that the number of seeds per ear in maize has increased about two fold in plants inoculated with *Azospirillum*. The interaction involving mycorrhiza *Azospirillum* and NP fertilizer was not significant for most of the yield and quality attributes studied. This is likely to be due to the interference of environmental factors on the tripartite symbiosis. However offering such comments require more in depth studies.

9. 4. Plant nutrient status

Nutrient status (N, P) of sesame plant was significantly improved due to inoculation with AM fungi (Table 9. 5) (Islam, 1980; Hawkins et al., 2000). Highest N and P contents were achieved under moderate to higher level of added N and P. However, the difference has not reached a significant level. *Azospirillum* alone as well as in combination with AM fungi and NP fertilizer (M×F×A) significantly improved the N and P status of plants. This observation could be ascribable to the collective efforts of both mycorrhiza and *Azospirillum*. AM fungi are known to increase P uptake in the host plant which in turn stimulate nitrogen uptake (Rao et al., 1985). Positive interaction between AM fungi and *Azospirillum* on plant nutrition has been described by several researchers (Barea et al., 1983; Rao et al., 1985; Tilak, 1995).

Table. 9. 5. Tissue N and P content as influenced by inoculation of AM fungi and *Azospirillum* under different fertilizer level in sesame.

AM level	Azo. level	Fert. level	N (mg g^{-1})		P (mg g^{-1})	
			35 DAS	70 DAS	35 DAS	70 DAS
M_0	A_0	0	26.00	25.31	11.10	12.09
		1	19.07	18.43	03.81	06.35
		2	26.29	11.32	05.45	07.05
		3	20.51	26.40	10.34	05.22
		4	30.83	34.06	06.26	08.12
	A_1	0	14.09	14.18	10.27	11.08
		1	12.07	15.99	11.05	10.96
		2	15.31	14.87	10.61	12.09
		3	19.78	16.70	08.39	08.75
		4	26.35	34.26	05.57	10.14
M_1	A_0	0	18.17	20.40	05.75	07.37
		1	23.75	26.30	10.70	07.84
		2	22.77	18.60	11.60	11.36
		3	24.48	22.79	09.21	12.73
		4	27.87	32.42	07.78	09.53
	A_1	0	24.14	24.69	10.50	10.80
		1	25.77	20.74	08.75	08.46
		2	12.65	24.83	11.08	11.33
		3	21.97	24.89	09.73	08.77
		4	13.02	17.73	13.85	16.61
Significance						
M			<0.006**		<0.006**	
F			<0.001***		0.288^{NS}	
A			<0.001***		<0.001***	
D			0.036*		0.151^{NS}	
M × F			<0.001***		<0.001***	
M × A			0.064^{NS}		0.485^{NS}	
M × D			<0.001***		0.755^{NS}	
M × F × A			<0.001***		<0.001***	
M × F × D			0.694^{NS}		<0.337^{NS}	
M × A × D			0.001***		0.492^{NS}	
M × F × A × D			0.843^{NS}		0.224^{NS}	

9. 5. Conclusion

The present study examined the individual and interactive effects of AM fungi, *Azospirillum* and NP fertilizers on mycorrhizal colonization, growth, yield and nutrient status of sesame. In general, AM colonization improved all the measured parameters under moderate levels of NP fertilizers. The inoculation of *Azospirillum* alone holds good only in the case of mycorrhizal colonization and nutrition of the crop. The interaction of AM fungi *Azospirillum* and NP fertilizer led to a general increment in AM colonization, growth and nutrient status of the crop. It can be concluded that the beneficial effect of the tripartite symbiosis on the growth and yield of sesame is not direct but through an enhanced uptake of nutrients.

CHAPTER 10

Efficacy of Different AM Inoculation Methods on Field Grown Sesame

10. 1. AM colonisation

Mycorrhizal parameters such as frequency of colonization (%F) and intensity of colonization (%M) in field grown sesame recorded a higher value in inoculated treatments, irrespective of the method by which the inoculum is introduced (Table 10.1) indicating that field inoculation with AM fungi greatly improves the mycorrhizal status of field grown plants (Mohammad et al., 1998). Among the various methods of inoculation, the plants received inoculation by SCAM method had the highest % F and % M possibly because of the reason that this method allows a close retention of inoculum within the immediate vicinity of germinating seeds so that the radicle can pick up the desired fungal infection. If the distribution of mycorrhizal inoculum is spatially patchy, it could contribute to variable rates of AM colonization among plants (Janos 1992; Lovelock and Miller, 2002). Lack of inoculum distribution may likely to have contributed to the low colonization by AM fungi observed under other inoculation methods. Hayman et al. (1981) observed a better AM colonization (30%) in field grown red clover under multi-seeded pelleting (seeds stuck on to pellets of soil inoculum) over applying the soil inoculum and seed broadcast, which brought only 10% of infection. Similarly Vassilev et al. (2001) obtained a root colonisation of 32 ± 5.6 and 24 ± 12.1 in tomato by application of gel entrapped spores of *G. deserticola* and *G. fasciculatum* respectively to field soil.

Table 10. 1. Efficacy of different AM inoculation methods on AM colonization in sesame

Treatment	Colonization	
	% F	% M
T_1 CON	18.89[c]	2.91[c]
T_2 IBC	38.33[b]	5.56[bc]
T_3 IBCLD	42.22[ab]	6.53[b]
T_4 SCAM	47.40[a]	11.40[a]
T_5 GEAM	36.48[b]	10.50[a]
T_6 MYCKE	40.56[ab]	5.33b[c]

Means in each column with different letters are significantly different ($P<0.05$) by Tukey's HSD

CON – Control (uninoculated)
IBC – Inoculum broadcast; IBCLD-Inoculum broadcast followed by land disturbance
SCAM – Seed coating with AM fungi; GEAM – Gel entrapped AM fungi
MYCKE- Mycorrhizal cakes

% F gradually increased with advancing growth stages (Fig. 10.1.) probably because of getting the newly emerged roots getting infected by AM fungi. Allen et al (1998) noticed that active root growth provided more entry points for AM fungi However the %M did not show any apparent increase with plant age (Fig. 10. 2) which coincides with the finding of Wilson and Tommerup (1992) and Pattinson and McGee (1997) that the proportion of colonized roots increases exponentially at first and then reaches a plateau.

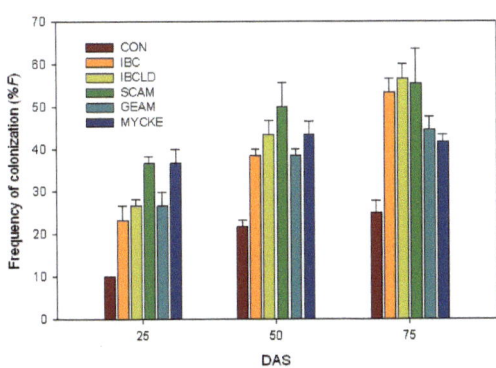

Fig. 10. 1 Efficacy of different AM inoculation methods on frequency of colonization (% F) during the growth stages of sesame.

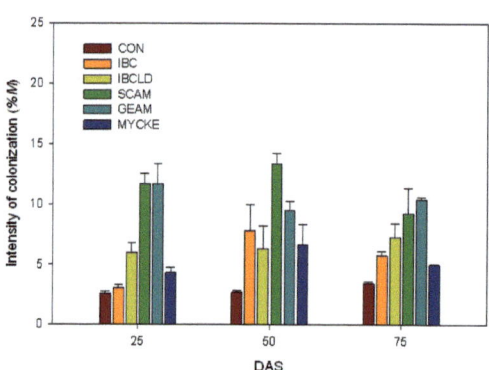

Fig. 10.2 Efficacy of different AM inoculation methods on intensity of colonization (%M) during the growth stages of sesame

10. 2. Growth characteristics

Irrespective of the methods, AM inoculation improved the growth characteristics of sesame such as root length, shoot length, leaf number and leaf area (Table 10.2). This could be due to the establishment of an effective mycorrhizal system with the formation of more extramatrical hyphae (Sanders and Tinker, 1971) resulting in a better harvest of immobile elements such as P (Sylvia et al 1993) which

Table 10. 2. Efficacy of different AM inoculation methods on growth characteristics of sesame

Treatment	Root length (cm)	Shoot length (cm)	Leaf no. plant^{-1}	Leaf area (cm^{-2} plant^{-1})
T_1	3.44^b	9.33^c	8.11^b	57.13^b
T_2	6.64^a	15.44^{ab}	10.44^{ab}	73.51^{ab}
T_3	7.63^a	12.50^{bc}	10.33^{ab}	72.12^{ab}
T_4	7.11^a	18.06^a	14.78^a	104.84^a
T_5	6.89^a	17.28^a	16.00^a	112.99^a
T_6	6.61^a	14.50^{ab}	10.67^{ab}	70.16^{ab}

Means in each column with different letters are significantly different ($P<0.05$) by Tukey's HSD

might have translated to growth. In the case of root length, the inoculation methods did not show much difference in its efficiency as the values were on par. Nevertheless, more shoot length, leaf number and leaf area were observed in SCAM and GEAM method of AM inoculation once again indicating that a close availability of AM inoculum with the emerging roots can hasten the AM infectivity and establishment which has a bearing on growth. The growth characteristics varied significantly with plant age, which showed a relatively high value at harvest (Fig 10. 3 – 6). Enhancement of plant growth following AM fungal inoculation has been demonstrated by several workers (Al-Karaki and Clark, 1998; Stewart et al., 2005; Idnani and Singh, 2008).

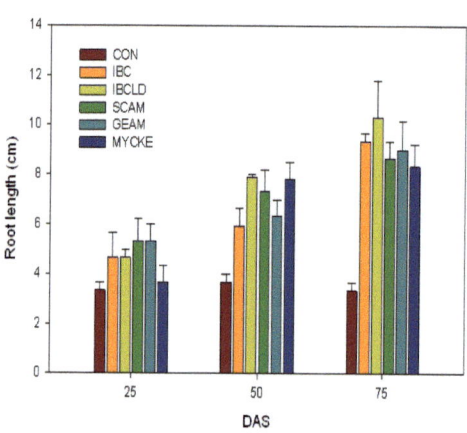

Fig. 10. 3 Efficacy of different AM inoculation methods on plant root length during the growth stages of sesame

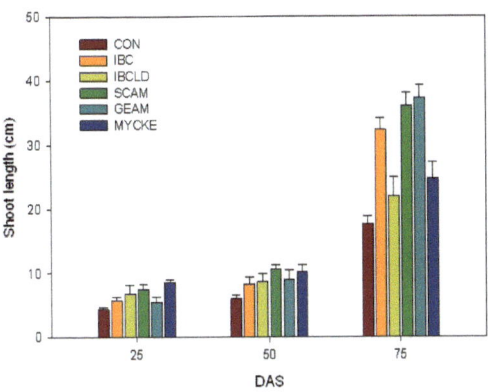

Fig. 10. 4. Efficacy of different AM inoculation methods on plant shoot length during the growth stages of sesame

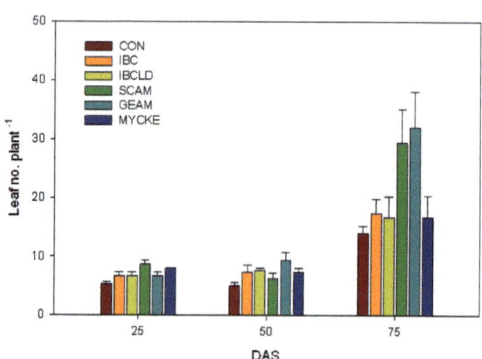

Fig. 10. 5. Efficacy of different AM inoculation methods on plant leaf number during the growth stages of sesame

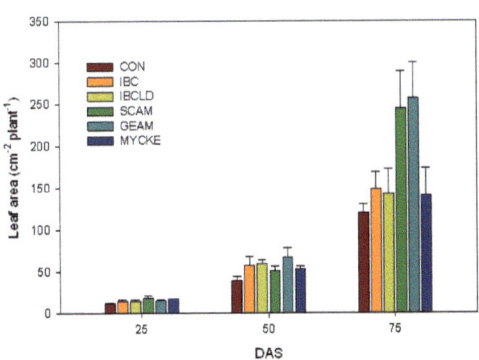

Fig. 10. 6. Efficacy of different AM inoculation methods on plant leaf area during the growth stages of sesame

10. 3. Yield attributes

In general, the AM inoculated plants promoted better yield in sesame irrespective of the method by which inoculum is applied. Most of the yield parameters such as pod number and seed number were significantly increased in GEAM method of AM inoculation while seed oil content was more under SCAM method (Table 10.3). Seed weight did not respond positively to any of the inoculation methods.

Table 10. 3. Efficacy of different AM inoculation methods on yield attributes of sesame

Treatment	Pod no plant^{-1}	Seed no plant^{-1}	Seed wt (g plant^{-1})	Oil content (%)
T_1	0.78^b	6.67^c	0.02^a	8.22^b
T_2	1.00^b	27.11^{abc}	0.08^a	8.88^b
T_3	1.33^b	8.55^{bc}	0.07^a	20.0^a
T_4	1.56^b	33.33^{ab}	0.10^a	22.22^a
T_5	2.89^a	47.33^a	0.11^a	13.28^b
T_6	1.11^b	27.11^{abc}	0.08^a	13.74^b

Means in each column with different letters are significantly different ($P<0.05$) by Tukey's HSD

The efficacy of inoculation methods at different stages of growth could not be compared in the present study as the plant enters into the reproductive stage only after 50 DAS (Fig. 10.7 – 10). Earlier reports indicate that barring the differences of each AM inoculation method in its efficiency, most of them help the field crops in realizing better yield (Bagyaraj, 1992).

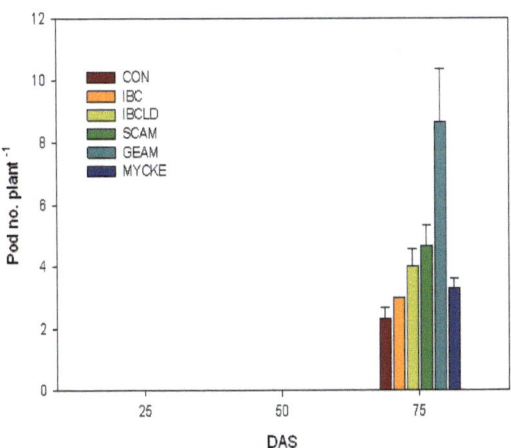

Fig. 10.7. Pod number in sesame at harvest stage under different AM inoculation methods

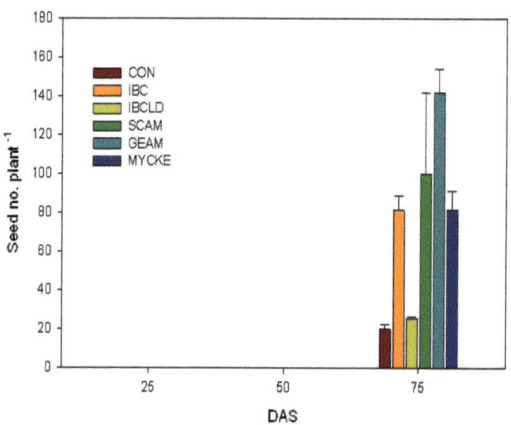

Fig. 10. 8. Seed number in sesame at harvest stage under different AM inoculation methods

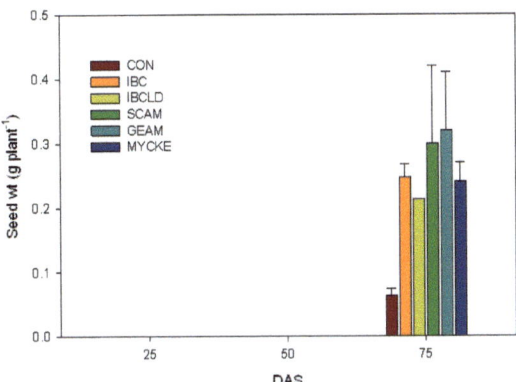

Fig. 10. 9. Seed weight in sesame at harvest stage under different AM inoculation methods

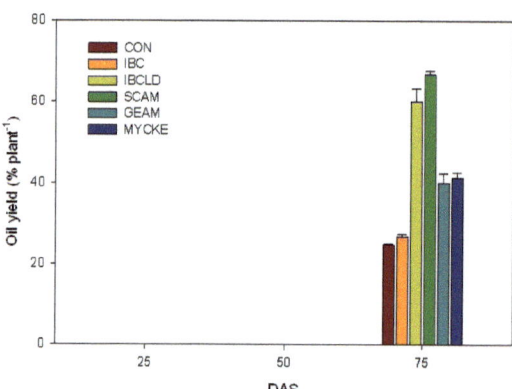

Fig. 10. 10. Oil content in sesame seeds at harvest stage under different AM inoculation methods

10. 4. Conclusion

Several methods for the introduction AM inoculum to field crops are in practice. However, none of the methods have been tried in sesame. The present study was therefore envisaged to fill in this gap. The study brought to light that the broadcast of seeds coated with AM inoculum and gel entrapped spores to field seems to be most effective over other methods in improving mycorrhizal colonization, growth and yield of the crop.

CHAPTER 11

General Discussion and Conclusion

Sesame is an important oilseed crop grown in Kerala for its edible seeds. It is mainly grown in sandy soils having poor fertility. Since the crop is grown as an interim crop in rice fallows, farmers always think about making profitable returns without any extra input of chemical fertilizers. In this situation the potential to increase the sustainability of sesame production through greater fertilizer use efficiency is possible by inoculating sesame seedlings with AM fungi.

AM colonization in sesame has already been reported. In order to make use of the mycorrhizal technology for the betterment of the crop, it is required to study the distribution pattern of the fungus in different soils and factors influencing their establishment in the host. Further, the efficient AM species/strains that enhance growth and nutrient uptake of the crop have to be identified. Since sesame is grown in different soil types, an understanding of the infective pattern of AM species under these soils enables to select the suitable and infective ones for inoculation schedules. Investigations on genotype dependent variation in AM colonization will lead to a better understanding of the host-microbe interaction. Sesame though, grown as a rainfed crop, occasional irrigation and application of chemical fertilizers are reported to increase productivity (Anonymous, 2007). A better understanding of the influence of agronomic practices and inputs is of utmost importance for the effective manipulation of the fungus. Additionally, the introduction of N_2 fixing organisms to the mycorrhizal system may further enhance the symbiotic efficiency. AM fungi being an obligate symbiont, it is worthwhile to identify a field inoculation method to optimize the production of sesame.

The survey, conducted during the first phase of the investigation revealed that there exists great variation in the extent of root colonization and spore density in the rhizosphere soil samples of sesame. Soil characteristics had a profound influence on both root colonization and diversity in the soil. The genus *Glomus* with its species *G.*

dimorphicum was the predominant one both in frequency and relative abundance. Previous studies (Selvaraj and Subramanian, 1988), witnessed the incidence of AM species such as *G. macrocarpum* and *S. sinosa* in sesame soils. However, those species were not encountered in the present study. Variably in root infectivity and spore density in root region soils of field crops has been reported earlier (Sulochana and Manoharachary, 1989; Valsalakumar et al., 2007).

The developmental pattern of fungal structures such as intercellular hyphae, vesicles and arbuscules differed by species of AM fungi infecting sesame with a more intensive production of these structures during the early stages of growth and establishment. It is speculated that the intense development of AM fungal structures during early stages is to meet the increasing demand of water and nutrients during vigorous vegetative growth. The importance of AM infectivity during the early stages of plant growth has been explained by Mohammad et al. (1998).

AM fungi differ in their ability to form efficient colonization with different crop plants. Evaluation of five AM species common in the rhizosphere of sesame exhibited differential response in augmenting mycorrhizal colonization and plant growth characteristics. For examples, the highest root colonization was offered to plants by the AM species *A. delicata*. On the contrary *A. lacunosa* enhanced maximum growth. However, considering the capability of *G. dimorphicum* in enhancing most of the parameters tested, it has been evaluated as the ideal AM species for sesame crop grown in this area. As mentioned earlier the occurrence of different AM species in the rhizosphere of sesame has been reported (Selvaraj and Subramanian, 1988). Nevertheless, the screening for the identification of an efficient species/strain for the crop has not been attempted.

It is a known fact that the AM population greatly fluctuates under different soil types. However the question as to how long the spores of AM fungi can remain incubated in the soil without losing its viability and infectivity still remains as a concern, which is not properly addressed. Irrespective of AM species, its infectivity on sesame declined with prolonged soil incubation, though the infectivity lasted up to five months. The infectivity of AM fungi on the crop was further altered in the availability of P. The study thus brings to light that the viability and infectivity of AM species is affected under prolonged incubation in all soil types in the absence of the host.

The extent of AM colonization in roots and the response to inoculation with AM fungi varies with plant species. Different cultivars of a single plant species also differ in their ability to harbor AM fungi in their root system. Sesame accessions screened in the present study showed variation to accommodate the fungi in their root system. Out of the twenty accessions only 30 % had a relatively high rate of infection. Similar genotype dependent variation has been reported by several workers. (Thomas and Ghai, 1987; Harikumar and Potty, 2002). Despite the low level of colonization, the crop was greatly dependent on indigenous AM fungi for growth and nutrition.

Soil management practices and disadvantageous crop rotation negatively affect AM fungi. Conversely, mycorrhizal fungi are usually more abundant in sustainable and medium-to-low input production systems. In order to make use of the benefit of AM fungi to the maximum extent, appropriate management practices have to be employed. In sesame, AM inoculation benefited the crop at the minimal input of P fertilization and irrigation. Increased crop productivity at a low level of fertilizer application and management practices is one of the most important beneficial effects offered by mycorrhizal symbiosis (Bolan et al., 1984; Harikumar and Thomas, 1991).

Dual inoculation with AM fungi and N_2 fixing endophytes may provide enough P and N to enhance plant growth in marginal environments. In sesame, AM colonization improved all the measured parameters under moderate levels of NP fertilizers. Inoculation of *Azospirillum* had a stimulatory effect on root colonization by AM fungi. Interaction of AM fungi, *Azospirillum* and NP fertilizers led to a general increment in AM colonization, growth and nutrient (N, P) status of the crop. The beneficial effect of tripartite symbiosis in crop production has been described by Barea et al. (1983) and Indi et al. (1990).

Mycorrhizal potential can be extended to crop plants only through an efficient delivery system to the site of their use. Various inoculation methods have been devised for field crops (Jackson et al., 1972; Harikumar and Potty, 2002). Among the five inoculation methods tried in the present investigation, broadcast of sesame seeds coated with AM inoculum and gel entrapped spores to the field were found to be the most effective one over other methods in enhancing the mycorrhizal colonization and general growth of the crop.

It is hoped that the understanding of the multifarious activities of AM fungi is likely to provide a prospective tool for sustainable production of sesame in different agroecosystems.

References

Abbot, L.K. and Robson, A.D. 1985. Formation of external hyphae in soil by four species of vesicular – arbuscular mycorrhizal fungi. *New Phytol.* **99**: 245 – 255.

Abbot, L.K. and Robson, A.D. 1985b. The effect of soil pH on the formation VA mycorrhizas by two species of *Glomus*. *Australian J. Soil Res.* **23**: 253-261.

Abbot, L.K. and Robson, A.D. 1991. Factors influencing the occurrence of vesicular – arbuscular mycorrhizas. *Agri. Ecosyst. Environ.* **35**: 121 – 150.

Agarwal, S.K. 2005. *Biofertilizers*. Advanced Environmental Biotechnology. APH Publishing Corporation, New Delhi

Agbenin, J.O. and Goladi, J.T. 1997. Carbon nitrogen and phosphorus dynamics under continuous cultivation as influenced by farmyard manure and inorganic fertilizers in the savanna of northern Nigeria. *Agric. Ecosys. Environ.* **63**: 17 – 24.

Aguilera – Go'mez, L., Davis, F.T. Jr, Olalde – Portugal, V., Duray, S.A. and Phavaphutanon, L. 1999. Influence of phosphorus and endomycorrhiza *(Glomus intraradices)* on gas exchange and plant growth of chile ancho pepper (*Capsicum annuum* L. CV San Luis) *Photosynthetica.* **36**: 441 – 449.

Ahiabor, B.D. and Hirata, H. 1994. Characteristic responses of three tropical legumes to the inoculation of two species of VAM fungi in Andosol soils with different fertilities. *Mycorrhiza* **5**: 63 – 70.

Al-Agely, A.K. and Reeves, F.B. 1995. Inland sand dune mycorrhizae: effects of soil depth, moisture and pH on colonization of *Oryzopsis hymenoides*. *Mycologia.* **87**: 54 – 60.

Alarcón, C. and Cuenca, G. 2005. Arbuscular mycorrhizas in coastal sand dunes of the Paraguaná Peninsula, Venezuela. *Mycorrhiza.* **16**: 1-9.

Alexander, I.J. and Fairley, R.I. 1983. Effects of nitrogen fertilization on population of feni roots and mycorrhizas in spruce humus. *Plant Soil.* **71**: 49 - 53.

Alexander, I.J. and Fairley, R.J. 1986. Growth and nitrogen uptake rates of ectomycorrhizae spruce seedlings. In: Gianinnazi – Pearson, V. and Gianinazzi, S (Eds.) Physiological and Genetical aspects of Mycorrhizae. Proceedings of 1st European symposium on Mycorrhiza. Dijon. INRA. Paris. pp. 377 – 382.

Al-Karaki, G.N. and Clark, R.B. 1998. Growth mineral acquisition and water use by mycorrhizal wheat grown under water stress. *J. Plant Nutr.* **21**: 263 – 276.

Allen, E.B., Rincon, E., Allen. M.F., Perez – Jimenez, A. and Huante, P.1998. Distribution and seasonal dynamics of mycorrhizae in a tropical deciduous forest in Mexico. *Biotropica.* **30**: 261 – 274.

Allen, M.F. 1992. Mycorrhizal functioning: An integrative plant – fungal process. Chapman and Hall, New York, NY 1001 – 2291.

Amijee, F., Stribley, D.P. and Tinker, P.B. 1990. Soluble carbohydrates in roots of leek (*Allium porrum*) plants in relation to phosphorous supply and VA mycorrhiza. *Plant Soil.* **124**: 195 – 198.

Amijee, F., Tinker, P.B. and Stribley, D.P. 1989a. The development of endomycorrhizal root systems. VII. A detailed study of effects of soil phosphorus on colonization .*New Phytol.* **111**: 435 – 446.

Amijee, F., Tinker, P.B. and Stribley, D.P. 1989b. Effects of phosphorus on the morphology of VA mycorrhizal root system of leek (*Allium porrum* L.). *Plant Soil.* **119**: 334 – 336.

Ammani, K. and Rao, A.S. 1996. Effect of two arbuscular mycorrhizal fungi *Acaulospora spinosa* and *A. scrobiculata* on upland rice varieties. *Microbiol. Res.* **151**: 235 – 237.

An, Z.Q., Shen, T. and Wang, H.G. 1993. Mycorrhizal fungi in relation to growth and mineral nutrition of apple seedlings. *Sci. Hort.* **54**: 175 – 285.

Anderson, E.L., Millner, P.D. and Kunishi, H.M. 1987. Maize root length density and mycorrhizal infection as influenced by tillage and soil phosphorus. *J. Plant Nutrition.* **10**: 1349 – 1356.

Anderson, R.C., Hetrick, B.A.D. and Wilson, G.W.T. 1994. Mycorrhizal dependence of *Andropogon gerardii* and *Schizachyrium scoparium* in two prairie soils. *Am. Midl. Nat.* **132**: 366-376.

Anonymous, 2007. Package of Practices, Kerala Agricultural University, Thrissur.

Anonymous, 2009a. Directorate of economics and statistics, Ministry of agriculture, Government of India. http://dacnet.nic.in

Anonymous, 2009b. Area and production of important crops in Kerala, Agricultural Statistics, Department of Economics and Statistics, Kerala. pp.9.

Antunes, V. and Cardoso, E.J.B.N. 1991. Growth and nutrient status of citrus plants as influenced by mycorrhiza and phosphorus application. *Plant Soil.* **13**: 11 – 19.

AOCS, 1993. *Official Methods and Recommended Practices.* The American Oil Chemists Society Champaign II.

Arines, J., Palma, J.M. and Vilarino, A.1993. Comparison of protein patterns in non-mycorrhizal and vesicular – arbuscular mycorrhizal roots of red clover. *New Phytol.* **123**: 763 – 768.

Arines, J., Quirtela, M., Vilarino, A. and Palma, J.M. 1994. Protein patterns and superoxide dismutase activity in non-mycorrhizal and arbuscular mycorrhizal *Pisum sativum* L. plants. *Plant Soil* **166**: 37 – 45.

Artursson, V., Finlay, R.D. and Jansson, J.K. 2006. Interaction between arbuscular mycorrhizal fungi and bacteria and their potential for stimulating plant growth. *Envtl. Microbiol.* **8**: 1-10.

Asghari, H.A., Marschner, P., Smith, S.E. and Smith, F.A. 2005. Growth response of Atriplex nummularia to inoculation with arbuscular mycorrhizal at different salinity levels. *Plant Soil* **273**: 245-256.

Ashri, A. 1998. Sesame breeding. *Plant Breeding Res.* **16**: 179 – 228.

Asif, M., Khan, A.G., Khan, M.A. and Kuck, C. 1995. Growth response of wheat to sheared root vesicular arbuscular mycorrhizae inoculum under field conditions In: Adholeya A. and Singh, S. (Eds.) *Mycorhizae: biofertilizers for the future.* Tata Energy Research Institute, New Delhi pp. 548.

Asimi, S., Gianinazzi – Peason, V. and Gianinazzi, S. 1980. Influence of increasing soil phosphorus levels on interactions between vesicular – arbuscular mycorrhizae on *Rhizobium* in soybeans. *Can. J. Bot.* **58**: 2200 – 2205.

Azcón – Aguilar, C. and Barea, J.M. 1997. Applying mycorrhiza biotechnology to horticulture: significance and potentials. *Sci. Hort.* **68**: 1 – 24.

Azcón, P. and Ocampo, N.A. 1981. Factors affecting vesicular – arbuscular infection and mycorrhizal dependency of thirteen wheat cultivars. *New Phytol.* **87**: 677 – 685.

Azcon, R., Ruiz – Lozano, J.M. and Rodriguez, R. 2001. Differential contribution of arbuscular mycorrhizal fungi to plant nitrate (^{15}N) under increasing N supply to the soil. *Can. J. Bot.* **79**: 1175 – 1180.

Baby, U.I. and Rao, M.K. 1996. Influence of organic amendments on arbuscular mycorrhizal fungi in relation to rice sheath blight disease. *Mycorrhiza.* **6**: 201 – 206.

Bagyaraj, D.J. and Majunath, A. 1980. Response of crop plants to VA mycorrhizal inoculation in an unsterile Indian Soil. *New Phytol.* **85**: 33 – 36.

Bagyaraj, D.J. 1992. Vesicular – arbuscular mycorrluza: Application in Agriculture. In: Norris, J. R., Read, D.J and Varma, A.K. (Eds) Methods *in Microbiology. Techniques for the study of mycorrhiza.* **24**: 359 – 373.

Barea, J.M. 2000. Rhizosphere and mycorrhiza of field crops. In: Balázs, E., Galante, E., Lynch, J.M., Schepers, J.S., Toutant, J.P., Wemer, D., Werry, P.A Th. J. (Eds.) *Biologcal Resource Management Connecting Science and Policy.* Springer-Verlag, Berlin. pp. 82-92.

Barea, J.M., Azcón-Aguilar, C. and Azcon, R. 1987. Vesicular-arbuscular mycorrhiza improve both symbiotic N_2 fixation and N uptake from soil as assessed with 15$_N$ technique under field conditions. *New Phytol.* **106**: 717 – 725.

Barea, J.M., Bonis, A.F. and Olivares, J. 1983. Interactions between *Azospirillum* and VA-mycorrhiza and their effects on growth and nutrition of maize and rye grass. *Soil Biol. Biochem.* **15**: 705 – 709.

Bates, L.S., Waldren, R.P. and Teare, I.D. 1973. Rapid determination of free proline for water stress condition. *Plant Soil* **39**: 205 – 207.

Baydar, H., Turgut, I. and Turgut, K. 1999. Variation of certain characters and line selection for yield, oil, oleic and linoleic acids in the Turkish sesame *(Sesamum indicum* L) population. *Tr. J. Agric. For.* **23**: 431 – 441.

Bedigian, D. and Harlan, J.R. 1986. Evidence for cultivation of sesame in the ancient world. *Econ. Bot.* **40**: 137 – 154.

Bever, J.D., Schultz, P.A., Pringle, A. and Morton, J.B. 2001. Arbuscular mycorrhizal fungi: more diverse than meets the eye, and the ecological tale of why. *Bio Science.* **51**: 923 – 931.

Bhardwaj, S., Dudeja, S.S. and Khurana, A.L.1997. Distribution of vesicular – arbuscular fungi in the natural ecosystem. *Folia Microbiol.* **42**: 589 – 594.

Bindu, M.V. and Harikumar, V.S. 2008. Diversity pattern of arbuscular mycorrhizal fungi in some contaminated sites of Kerala. *Mycorrhiza News* **20**: 9-10.

Biro, B., Péchy, K.K., Vörös I., Takács, T., Eggenberger, P. and Strasser, R.J. 2000. Interaction between *Azospirillum* and *Rhizobium* nitrogen fixers and arbuscular mycorrhizal fungi in the rhizosphere of alfalfa in sterile AMF-free on normal soil conditions. *Applied Soil Ecol.* **15**: 159 – 168.

Black, C.A. 1965. Methods of soil analysis. *In: Agronomy Part II* Vol. 9. Am. Soc. Agron, Wisconsin, U.S.A., pp. 1114 – 1162.

Blal, B., Morel, C., Gianinazzi – Pearson, V., Fardeau, J.C. and Gianinazzi, S. 1990. Influence of vesicular – arbuscular mycorrhizae on phosphate fertilizer efficiency in two tropical soils planted with micropropagated oil palm *(Elaeis guineensis* jacq.). *Biol. Fertil. Soils.* **9**: 43 – 48.

Bolan, N.S., Robson, A.D. and Barrow, N.J. 1984. Increasing phosphorus supply can increase the infection of plant roots by vesicular – arbuscular mycorrhizal fungi. *Soil Biol. Biochem.* **16**: 419 – 420.

Boucher, A., Dalpé, Y. and Charest, C. 1999. Effect of arbuscular mycorrhizal colonization of four species of *Glomus* on physiological responses of maize. *J. Plant Nutr.* **22**: 783 – 797.

Boureima, S., Diouf, M., Diop, T.A., Diatta , M., Leye, E.M., Ndiaye, F. and Seck, D. 2007. Effects of arbuscular mycorrhizal inoculation on the growth and development of sesame (*Sesamum indicum* L.). *African J. Agric. Res.* **3**: 234 – 238.

Bradford, M.M. 1976. A rapid and sensitive method for the quantitation of microgram quantities of protein utilizing the principle of protein-dye binding. *Anal. Biochem.* **72**: 248 – 254.

Branzanti, B., Gianinazzi – Pearson, V. and Gianinazzi, S.1992. Influence of phosphate fertilization on the growth and nutrient status of micropropagated apple infected with endomycorrhizal fungi during the bearing stage. *Agronomie.* **12**: 841 – 845.

Bray. S.R., Kitajima, K. and Sylvia, D.M. 2003. Mycorrhizas differentially alter growth, physiology and competitive ability of an invasive shrub. *Ecol. Appl.* **13**: 565 – 574.

Bruce, A., Smith, S.E. and Tester, M. 1994. The development of mycorrhizal infection in cucumber. Effects of P supply on root growth, formation of entry points and growth of infection units. *New Phytol.* **127**: 507 – 514.

Bürkert, B. and Robson, A., 1994. ^{65}Zn uptake in subterranean clover (*Trifolium subterraneum* L.) by three vesicular – arbuscular mycorrhizal fungi in a root free sandy soil. *Soil Biol. Biochem.* **26**: 1117 – 1124.

Cade- Menun, B.J., Berch, S.M. and Bomke, A.A. 1991. Seasonal colonization of winter wheat in south coastal British Columbia by VAM fungi. *Can. J. Bot.* **69**: 78-86.

Caglar, S. and Bayram, A. 2006. Effects of vesicular – arbuscular fungi on the leaf nutritional status of four grape vine root stocks. *European J. Hort. Sci.* **7**: 109 – 113.

Callow, J.A., Capaccio, L.C.M., Parish, G. and Tinker, P.B. 1978. Detection and estimation of polyphosphate in vesicular – arbuscular mycorrhizae. *New Phytol.* **80**: 125 – 134.

Cerligione, L.J., Liberta, A.E. and Anderson, R.C. 1988. Effect of soil moisture and soil sterilization on VA mycorrhizal colonization and growth of little bluestem *(Schizachyrium scoparium)*. *Can .J. Bot.* **66**: 757 – 761.

Chiramel, T., Bagyaraj, D.J. and Patil C.S.P. 2006. Response of *Andrographis paniculata* to different arbuscular mycorrhizal fungi. *J. Agril. Technol.* **22**: 221 – 228.

Christensen, B.T.1988. Effects of animal manure and mineral fertilizers on the total carbon and nitrogen contents of soil size fractions. *Biol. Fertil. Soils.* **5**: 304 – 307.

Chung, J., Lee, J. and Choe, E. 2004. Oxidative stability of soybean and sesame oil mixture during frying of flour dough .*J. Food Sci.* **69**: 574 – 578.

Clark, F.B. 1963. Endotrophic mycorrhizae influence yellow poplar seedling growth. *Science.* **140**: 1220 – 1221.

Clark, R.B. 1983. Plant genotype differences in the uptake, translocation, accumulation and use of mineral elements required for plant growth. *Plant Soil.***72**: 175 – 196.

Clark, R.B., Zobel, R.W. and Zeto, S.K. 1999. Effects of mycorrhizal fungal isolates on mineral acquisition by *Panicum virgatum* in acidic soil. *Mycorrhiza* **9**: 167 – 176.

Cook, R.C. and Whipps, J.M. 1993. *Ecophysiology of Fungi.* Blackwell Scientific Publications, Oxford.

Cooper, K.M. and Tinker, P.B. 1981. Translocation and transfer of nutrients in vesicular -arbuscular mycorrhizas IV. Effect of environmental variables on movement of phosphorus. *New Phytol.* **88**: 327 – 339.

Cox, G. and Tinker, P.B. 1976. Translocation and transfer of nutrients in vesicular arbuscular mycorrhizas I. The arbuscule and phosphorus transfer: a quantitative ultra structural study. *New Phytol.* **77**: 271 – 378.

Daft, M.J. and El-Giahmi, A.A. 1976. Studies on nodulated and mycorrhizal peanuts. *Ann. Appl. Biol.* **83**: 273 – 276.

Daft, M.J. and Elgiahmi, A.A. 1976. Studies on nodulated and mycorrhizal peanuts. *Ann. Appl. Biol.* **83**: 273-276.

Daft, M.J. and Nicolson, T.H.1969a. Effect of *Endogone* mycorrhiza on plant growth. II Influence of soluble phosphate on endophyte and host in maize. *New Phytol.* **68**: 945 – 952.

Dalal, S. and Hippalgaonkar, K.V.1995. The occurrence of vesicular – arbuscular mycorrhizal fungi in arable soils of Konkan and Solapur. In: Adholeya, A. and Singh, S., *Mycorrhiza biofertilizers for the future*. Proceedings of the third National Conference on Mycorrhiza 13 – 15 March. Tata Energy Research Institute, New Delhi, India. pp. 1 – 7.

Dangeard, P.A. 1990. Le *Rhizophagus populinus* Dangeard. *Botaniste* **7**: 225-291.

Daniels, B.A. and Skipper, H.D. 1982. Methods for the recovery and quantitative estimation of propagates from soil. In: Schenck, N.C. (Ed.) *Methods and Principles of Mycorrhizal Research*. American Phytopathological Society St. Paul, Minn. pp. 29 – 35.

Danneberg, G., Latus, C., Zimmer, W., Hundeshagen, B., Schneider – Poetsch, H.J. and Bothe, H. 1992. Influence of vesicular – arbuscular mycorrhiza on phytohormone balances in maize (*Zea mays* L.). *J. Plant Physiol.* **141**: 33 – 39.

de Fransca, 1981. Differences in dry matter yield and the uptake distribution and use of nitrogen by *Sorghum* genotypes. Ph.D. Thesis, University of Nebraska, Lincoln, N.E.

de Miranda, J.C.C., Harris, P.J. and Wild, A. 1989. Effects of soil and plant phosphorus concentrations on vesicular – arbuscular mycorrhiza in sorghum plants. *New Phytol.* **112**: 405 – 410.

Declerck, S., Plenchette, C. and Strullu, D.G. 1995. Mycorrhizal dependency of banana (*Musa* acuminata, AAA group) cultivar. *Plant Soil*. **176**: 183 – 187.

Dehne, H.W. and Backhaus, G.F. 1986 The use of vesicular – arbuscular mycorrhizal fungi in plant production. I. Inoculum production. *J. Plant Dis. Protect.* **93**: 415 – 424.

Potty, V.P. 1990. Use of lignite slurry as inoculating medium for vesicular-arbuscular mycorrhiza in Chinese potato *(Coleus parviflorus). Plant Soil* **125**: 146 – 148.

DeMiranda, J.C.C., Marris, P.J. and Wild, A. 1989. Effect of soil and plant phosphorus concentrations on vesicular – arbuscular mycorrhizae in sorghum plant. *New Phytol.* **112**: 405 – 410.

Dhillion, S.S. 1992. Evidence for host - mycorrhizal preference in native grassland species. *Mycol Res.* **94**: 359-362.

Dhillion, S.S., Anderson, R.C. and Liberta, A.E. 1988. Effect of fire on the mycorrhizal ecology of little bluestem (*Schizachyrium scoparium*). *Can. J. Bot.* **66**: 706-713.

Diaza, G., Roldan, P. and Abbaladeje, J. 1992. Influence of soil type in colonization patterns and efficiency of mycorrhiza: symbiosis of six *Glomus* sp. *Mycologia* **13**: 47 – 56.

Diederichs, C. 1992. Impact of tropical VA mycorrhizae on growth promotion of *Cajanus_cajan* as influenced by P sources and P levels. In: Jasper, D. (Ed.) *Proc Intl. Sym. on Management of Mycorrhizas in Agriculture, Horticulture and Forestry.* Australian Institute of Agricultural Sciences 28 Sept – 2 Oct 1992, University of Western Australia, Nedlands Australia.

Diem, H.G., Jung, G., Mugnier, J., Ganry, F. and Dommergues, Y. 1981. Alginate entrapped *Glomus mosseae* for crop inoculation In: *Proceedings of the Fifth North American Conference on Mycorrhizae,* University of Laval Qubec, Canada.

Dinkelaker, B. and Marschner, H. 1992. *In vivo* demonstration of acid phosphatase activity in the rhizosphere of soil-grown plants. *Plant Soil* **144**: 199 – 205.

Diop, T.A., Krasova – Wade, T., Diallo, A., Diouf, M. and Gueye, M. 2003. *Solanum* cultivar response to arbuscular mycorrhizal fungi: growth and mineral status. *African J. Biotechnol.* **2**: 429 – 433.

Dixon, R.K., Gary, V.K. and Rao, M.V. 1993. Inoculation of *Leucaena* and *Prosopis* seedlings with *Glomus* and Rhizobium species in a saline soil: Rhizosphere relations and seedling growth. *Arid Soil Res. Rehab.* **7**: 133 – 144.

Dodd, J.C., Arias, I., Koomen, I. and Hayman, D.S. 1990. The management of populations of VAM fungi in acid infertile soil of a savanna system. *Plant Soil* **122**: 229 – 240.

Douds Jr, D.D., Galvez, L., Frankesnyder, M., Reider, C. and Drinkwater, L.E. 1997. Effect of compost addition and crop rotation point upon VAM fungi. *Agric Ecosys. Environ.* **65**: 257 – 266.

Douds, D.D., Galvez, L., Janke, R.R. and Wagoner, P. 1995. Effect of tillage and farming system upon population and distribution of vesicular – arbuscular mycorrhizal fungi. *Agric. Ecosys. and Environ.* **52**: 111 – 118.

Dumas, E., Gianinazzi – Pearson, V. and Gianinazzi, S. 1989. Production of new soluble proteins during VA endomycorrhiza formation. *Agric. Ecosys. Environ.* **29**: 111 – 114.

Dumas, E., Tahiri – Alaout, A., Gianinazzi, S. and Gianinazzi – Pearson, V. 1990. Observations on modifications in gene expression with VA endomycorrhiza development in tobacco: qualitative and quantitative changes in protein profiles. In: Nardon, P., Gianinazzi – Pearson, V., Grenier, A.M., Margulis, L., Smith, D.C. (Eds.) *Endocytobiology* IV. INRA Edn. pp. 153 – 157.

Eid, M.T., Black, C.A. and Kempthorine, O. 1951. Importance of soil organic and inorganic phosphorus on plant growth at low and high soil temperatures. *Soil Soc. Am. J.* **15**: 89.

Estaun, M.V. 1991. Effect of Noel NaCl mannitol on the germination of two isolates of the vesicular – arbuscular mycorrhizal fungus *Glomus mosseae*. *Proceedings Third European Symposium on Mycorrhizas*, Univ of Sheffield, Sheffield, U.K.

Estaun, V., Calvet, C. and Hayman, D.S. 1987. Influence of plant genotype on mycorrhizal infection: Response of three pea cultivars. *Plant Soil*. **103**: 295 – 298.

Estrada-Luna, A.A. and Davis, F.T. 2003. Arbuscular mycorrhizal fungi influence water relations, gas exchange abscisic acid and growth of micropropagated chile ancho pepper (*Capsicum annum*) plantlets during acclimatization and post acclimatization. *J. Plant Physiol.* **160**: 1073 – 1083.

Evans, D.G. and Miller, M.H. 1990. The role of the external mycelial network in the effect of soil disturbance upon vesicular – arbuscular mycorrhizal colonization of maize. *New Phytol.* **114**: 65 – 71.

Fairchild, G.L. and Miller, M.H. 1988. Vesicular – arbuscular mycorrhizas and the soil disturbance – induced reduction of nutrient absorption in maize II. Development of the effect. *New Phytol.* **110**: 75 – 84.

Fairchild, G.L. and Miller, M.H. 1990. Vesicular – arbuscular mycorrhizas and the soil disturbance induced reduction of nutrient absorption in maize III. Influence of P amendments to soil. *New Phytol.* **114**: 641 – 650.

Faithfull, N.T. 2002. *Methods in Agricultural Chemical Analysis: A Practical Handbook.* CABI Publishing, Wallingford, U.K.

Ferguson, J.J. 1981. Inoculum production and field application of vesicular – arbuscular mycorrhizal fungi. Ph.D. Dissertation, University of California, Riverside, pp.117.

Fortuna, P., Citernesi, S., Morini, S., Giovannetti, M. and Loreti, F.1992. Infectivity and effectiveness of different species of arbuscular mycorrhizal fungi in micropropagated plants of Mr S2/5 Plum root stock. *Agronomie.* **12**: 825 – 829.

Frey, and Schuepps, H. 1993. Acquisition of nitrogen by external hyphae of arbuscular mycorrhizal fungi associated with *Zea mays* L. *New Phytol.* **124**: 221 – 230.

Fulchieri, M. and Frioni, L. 1994. *Azospirillum* inoculation on maize (*Zea mays*): Effect on yield in a field experiment in Central Argentina. *Soil. Biol. Biochem.* **26**: 921 – 923.

Furlan, V. and Bernier – Cardou, M. 1989. Effects of N, P and K on formation of vesicular – arbuscular mycorrhizae, growth and mineral content of onion. *Plant Soil* **113**: 167 – 174.

Ganesan, V. and Mahadevan, A. 1994. Effect of mycorrhizal inoculation of cassava, elephant foot yarn and taro. *J. Root Crops* **20**: 1 – 14.

Gaonker, S.B.N. and Sreenivasa, M.N. 1994. Effects of inoculation with *Glomus fasciculatum* in conjunction with different organic amendments on growth and yield of wheat (*Triticum aestivum* L.). *Microbiol. Res.* **149**: 419 – 423.

Gardner, W.H. 1986. Water content In: Klute, A (Ed.) *Methods of Soil Analysis Part I: Physical and Mineralogical Methods.* Second ed. America Society of Agronomy, Inc. Madison, W.I. pp. 493 – 544.

Gaur, A., Adholeya, A. and Mukerji, K.G. 1998. A comparison of AM fungi inoculants using *Capsicum* and *Polianthes* in marginal soil amended with organic matter. *Mycorrhiza* **7**: 307 – 312.

Gemma, J.N., Koske, R.E. and Carreiro, M. 1989. Seasonal dynamics of selected species of VA mycorrhizal fungi in a sand dune. *Mycol. Res.* **92**: 317-321.

George, E., Hausser, K.U., Vetterlein, D., Gorgus, E. and Marschner, H. 1992. Water and nutrient translocation by hyphae of *Glomus mosseae*. *Can. J. Bot.* **70**: 2130 – 2137.

Gerdemann, J.W. 1968. Vesicular arbuscular mycorrhiza and plant growth. *Annu Rev. Phytopath.* 6: 397-418.

Gerdemann, J.W. and Nicolson, T.H.1963. Spores of mycorrhizal *Endogone* species extracted from soil by wet sieving and decanting. *Trans. Br. Mycol. Soc.* **46**: 235 – 244.

Gerdemann, J.W. and Trappe, J.M. 1974. The Endogonaceae in the Pacific North West. *Mycologia Memoir.* 5: 1-76.

Gianinazzi – Pearson, V. and Azcón – Aguilar, C. 1991. Fisiologia de las microrizas vesiculo-arbusculares. In: Olivares, J. and Barea, J.M. (Eds.) Fijacion Y movilizacion biologica denutrientes, Vol. II, CSIC, Madrid. pp. 175 – 202.

Gianinazzi, S. and Gianinazzi – Pearson, V., Dexheimer, J. 1979. Enzymatic studies on the metabolism of vesicular-arbuscular mycorrhiza. III. Ultrastructural localization of acid and alkaline phosphatase in onion roots infected by Glomus mosseae (Nicol. and Gerd.). New Phytol. **82**: 127 – 132.

Giovannetti, M. 2000. Spore germination and pre-symbiotic mycelial growth. In: Kapulnik, Y. and Douds, D.D. (Eds.) *Arbuscular Mycorrhizas: Physiology and Function.* Kluwer Academic Publishers, Dordrecht, The Netherlands.

Giovannetti, M. 2004. Survival strategies in arbuscular mycorrhizal symbiosis. In: Seckbach, J. (Ed.) *Symbiosis.* Kluwer Academic Publishers, Dordrecht. pp. 293.

Giovannetti, M. and Citernesi, A.S., 1993. Time-course of appresorium formation on host plants by arbuscular mycorrhizal fungi. *Mycol. Res.* **97**: 1140 – 1142.

Giri, B. and Mukerji, K.G. 1999. Improved growth and productivity of *Sesbania grandiflora* pers under salinity stress through mycorrhizal technology. *J. Phytological* Res. **12**: 35 – 38.

Girija, V.K. and Nair, S.K. 1985. Occurrence of vesicular arbuscular mycorrhiza in certain crop plants of Kerala. *Agric. Res. J. Kerala.* **23**. 185 – 188.

Girija, V.K. and Nair, S.K. 1985. Occurrence of vesicular arbuscular mycorrhiza in certain crop plants of Kerala. *Agric. Res. J. Kerala.* **23**: 185 – 188.

Goicoechea, N., Szalai, G., Antolin, M.C., Sanchez Diaz, M. and Paldi, E. 1998. Influence of arbuscular mycorrhizae and *rhizobium* on free polyamines and proline levels in water stressed alfalfa. *Hort Sci.* **33**: 706 – 711.

Graham, J. H. and Eissenstat, D.M. 1994. Host genotype and the formation and function of VA mycorrhizae. *Plant Soil.* **159**: 179 – 185.

Graham, J.H., Duncan, L.W. and Eissenstat, D.M. 1997. Carbohydrate allocation pattern in citrus genotypes as affected by phosphorus nutrition, mycorrhizal colonization and mycorrhizal dependency. *New Phytol.* **135**: 335 – 343.

Graham, J.H., Eissenstat, D.M. and Drouillard, D.L. 1991. On the relationship between a plant's mycorrhizal dependency and rate of vesicular arbuscular mycorrhizal colonization. *Funct. Ecol.* **5**: 773 – 779.

Graham, J.H., Leonard, R.T. and Menge, J.A. 1981. Membrane mediated decrease in root exudation responsible for phosphorus inhibition of vesicular – arbuscular mycorrihza formation. *Plant Physiol.* **68**: 548 – 552.

Graham, J.H., Linderman, R.G. and Menge, J.A. 1982. Development of external by different isolates of mycorrhizal *Glomus* sp. in relation to root colonization and growth of *Troyer citrange. New Phytol.* **91**: 183 – 189.

Graw, D. 1979. The influence of soil pH on the efficiency of vesicular – arbuscular mycorrhiza. *New Phytol.* **82**: 687-695.

Grazey, C., Abbot, L.K. and Robson, A.D. 2004. Indigenous and introduced arbuscular mycorrhizal fungi contribute to plant growth in two agricultural soils from south-western Australia. *Mycorrhiza.* **14**: 355 – 362.

Gryndler, M., Vósatka, M., Hršelová, H., Chavatalová, I. and Jansa .J. 2002. Interaction between arbuscular mycorrhizal fungi and cellulose in growth substrate. *Appl. Soil. Ecol.* **19**: 279 – 288.

Gupta, A.K., Shruti, C. and Sharma, A.K. 2009. Arbuscular mycorrhizal fungal diversity in some medicinal plants. *Mycorrhiza News* **20**: 10-13.

Gupta, M.L., Prasad, A., Ram, M. and Kumar, S. 2002. Effect of vesicular arbuscular (VAM) fungus *Glomus fasciculatum* on the essential oil yield related characters and nutrient acquisition in the crops of different cultivars of menthol mint (*Mentha arvensis*) under field conditions. *Bioresour. Technol.* **81**: 177 – 179.

Gupta, R. and Krishnamurthy, K.V. 1996. Response of mycorrhizal and non-mycorrhizal *Arachis hypogaea* to NaCl and acid stress. *Mycorrhiza.* **4**: 197 – 200.

Gupta, S.B. 1996. Effective utilization of phosphorus in rice-wheat cropping system in vertisol through VA mycorrhizae and phosphorus solubilizer. Ph. D Thesis IARI, New Delhi, India.

Haas, J.H. and Krikun, J. 1985. Efficacy of endomycorrhizal fungus isolates and inoculum quantities required for growth response. *New Phytol.* **100**: 613-621.

Habte, M. and Aziz, T. 1985. Response of *Sesbania grandiflora* to inoculation of soil with vesicular – arbuscular fungi. *Appl. Environ. Microbiol.* **50**: 701 – 705.

Hall, I.R. 1979. Soil pellets to introduce vesicular – arbuscular mycorrhizal fungi into soil. *Soil Biol. Biochem.* **11**: 85 – 86.

Hall, I.R. 1980. Growth of *Lotus pedunculatus* Cav. in an eroded soil containing soil pellets infected with endomycorrhizal fungi. *New Zealand J. Agric. Res.* **23**:103 – 105.

Hamel, C., Furlan, V. and Smith, D.L. 1991. N_2 fixation and transfer in a field grown mycorrhizal corn and soybean intercrop. *Plant Soil.* **133**: 177 – 185.

Harekrishna, Singh, S.K. and Patel, V.B. 2006. Screening of arbuscular – mycorrhizal fungi for enhanced growth and survival of micropropagated grape *(Vitis vinifera) plantlets. Indian J. Agric. Sci..* **76**: 297 – 301.

Harikumar, V.S. and Potty, V.P. 1999. Diversity pattern of endomycorrhizal association with sweet potato in Kerala. *J. Mycol. Pl. Pathol.* **29**: 197 – 200.

Harikumar, V.S. and Potty, V.P. 2002. Genotype dependent variation in arbuscular mycorrhizal colonization and its response on the growth of sweet potato. *Bulgarian J. Agric. Sci.* **8**: 161 – 166.

Harikumar, V.S. and Potty, V.P. 2002. Technology for mass multiplication of AMF for field inoculation to sweet potato. *Mycorrhiza News* **14**: 11 – 12.

Harikumar, V.S. and Thomas, G.V. 1991. Effect of fertilizer and irrigation on vesicular arbuscular mycorrhizal association in coconut. *Philippine J. Coconut Studies.* **16**:20 – 24.

Harinikumar, K.M. and Bagyaraj, D.J. 1989. Effect of cropping sequence, fertilizers and farm-yard manure on vesicular – arbuscular mycorrhizal fungi in different crops over three consecutive seasons. *Biol. Fertil. Soils* **7**: 173 – 175.

Hawkins, H.J., Johansen, A. and George, E. 2000. Uptake and transport of organic and inorganic nitrogen by arbuscular mycorrhizal fungi. *Plant Soil* **226**: 275 – 285.

Hayman, D.S. 1970. *Endogone* spore numbers in soil and vesicular arbuscular mycorrhizae as influenced by season and soil treatment. *Trans. Br. Mycol. Soc.* **54**: 53 – 63.

Hayman, D.S. 1973. Plant growth response to vesicular arbuscular mycorrhiza VI. Effect of light and temperature. *New Phytol.* **73**: 71 – 80.

Hayman, D.S. 1975. The occurrence of mycorrhiza in crops as affected by soil fertility. In: Sanders, F.E, Mosse, B. and Tinker, P.B. (Eds.) *Endomycorhizas* Academic Press, London, pp. 495 – 509.

Hayman, D.S. 1983. The physiology of vesicular arbuscular endomycorrhizae symbiosis. *Can. J. Bot.* **61**: 944 – 963.

Hayman, D.S., Barea, J.M. and Azcon, R. 1976. Vesicular arbuscular mycorrhiza in southern Spain. Its distribution in crops growing in soils of different fertility. Phytopathol Mediterranea. **15**: 1-6.

Hayman, D.S. and Mosse, B. 1979. Improved growth of white clover in hill grass lands by mycorrhizal inoculation. *Ann. Appl. Biol.* **93**: 141 – 148.

Hayman, D.S. and Tavares, M. 1985. Plant growth responses to vesicular – arbuscular mycorrhiza XV. Influence of soil pH on the symbiotic efficiency of different endophytes. *New Phytol.* . **100**: 367-377.

Hayman, D.S., Morris, E.J. and Page, R.J. 1981. Methods for inoculating field crops with mycorrhizal fungi. *Ann. Appl. Biol.* **99**: 247 – 253.

Hedge, J.E. and Hofreiter, B.T.1962. Estimation of carbohydrate In: Whistler, R.L. and Be Miller, J.N. (Eds.) *Carbohydrate Chemistry*, Academic Press, New York.

Helgason, T., Daniell, T.J., Husband, R., Fitter, A.H. and Young, J.P.W. 1998. Ploughing up the wood-wide web? *Nature* **394**: 431.

Hibasami, H., Fujikawa, T., Takeda, H., Nishibe, S., Satoh, T., Fujisawa, T. and Nakashima, K.2000. Induction of apoptosis by *Acanthopanax senticosus* HARMS and its component, sesamin in human stomach cancer KATO III cells. *Oncol. Rep.* **7**: 1213 – 1216.

Hildebrandt, U., Ouzaid, F., Marner, F.J. and Bothe, H. 2006. The bacterium *Paenibacillus validus* stimulates growth of the arbuscular, mycorrhizal fungus *Glomus intraradices* up to the formation of fertile spores. *FEMS Microbiol. Lett.* **254**: 258 – 267.

Hirrel, M.C. and Gerdemann, J.W. 1980. Improved growth of onion and bell pepper in saline soils by two vesicular – arbuscular mycorrhizal fungi. *Soil Sci. Soc. Am. J.* **44**: 654 – 655.

Ho, I. 1987. Vesicular – arbuscular mycorrhizae of halophyte grasses in the Alvard desert of Oregon. *Northwest Sci.* **61**: 148 – 151.

Hrśelová, H., Chavatalová, I., Vosátka, M., Klír, J. and Gryndler, M. 1999. Correlation of abundance of arbuscular mycorrhizal fungi, bacteria and saprophytic microfungi with soil carbon, nitrogen and phosphorus. *Folia Microbiol* **44**: 683 – 687.

Huang, R.S., Smith, W.K. and Yost, R.S. 1985. Influence of vesicular – arbuscular mycorrhiza on growth, water relations, leaf orientation in *Leucaena leucocephala* (Lam.) de wit. *New Phytol.* **99**: 229 – 243.

Husband, R., Herre, E.A. and Young, J.P.W. 2006. Temporal variation in the arbuscular mycorrhizal communities colonizing seedlings in a tropical forest. *FEMS Microbiol. Ecol.* **42**: 131 – 136.

Idnani, L.K. and Singh, R.J. 2008. Effect of irrigation regimes, planting and irrigation methods and arbuscular mycorrhizae on productivity, nutrient uptake and water use in summer green gram (*Vigna radiata* var *radiata*) *Indian J. Agric. Sci.* **78**: 53 – 57.

Indi, D.V., Konde, B.K., Wani, P.V. and Kale, P.N. 1990. Yield and nutrient uptake by brinjal as influenced by *Azospirillum brasilense* and/or *Glomus fasciculatum* inoculation under graded phosphorus levels. In: Jalali, B.L. and Chand, H. (Eds). *Current Trends in Mycorrhizal Research*. Proceedings of the National Conference on Mycorrhiza, Haryana Agricultural University, Hisar, India 14 – 16 February 1990, New Delhi. TERI. pp. 157 – 159.

Ishii, T., Narutaki, A., Sawada, K., Aikawa, J., Matsumoto, I. and Kadoya, K. 1997. Growth stimulatory substances for vesicular – arbuscular mycorrhizal fungi in Bahia grass (*Paspalum notatum* Flügge) roots. In: Ando, T., Fujita, K., Mae, T., Matsumoto, H., Mori, S. and Sekiya, J. (Eds.). *Plant nutrition for sustainable food production and environment.* Kluwer, Dordrecht, The Netherlands. pp. 733 – 736.

Ishii, T., Shrestha, Y., Matsumoto, I. and Kadoya, K. 1996. Effects of ethylene on growth of vesicular – arbuscular mycorrhizal fungi and on the mycorrhizal formation of trifoliate orange roots. *J. Jap. Soc. Hort. Sci.* **65**: 525 – 529.

Islam, R., Ayanaba, A. and Sanders, F.E. 1980. Response of cowpea (*Vigna unguiculata*) to inoculation with VA mycorrhizal fungi and rock phosphate fertilization in some unsterile Nigerian soils. *Plant Soil.* **54**: 107 – 117.

Jackson, M.L. 1973. *Soil Chemical Analysis*, Prentice Hall, New Delhi.

Jackson, N.E., Franklin, R.E. and Miller, R.H. 1972. Effects of vesicular – arbuscular mycorrhizae on growth and phosphorus content of three agronomic crops. *Soil Sci. Amer. Proc.* **36**: 64 – 67.

Jakobsen, I. 1987. Effects of VA mycorrhiza on yield and harvest index of field grown pea. *Plant Soil.* **98**: 407 – 415.

Jakobsen, I. and Nielson, N.E., 1983. VA mycorrhiza in field- grown crops. I. Mycorrhizal infection in cereals and peas at various times and soil depths. *New Phytol.* **93**: 401-413.

Jakobsen, I., Abbot, L.K. and Robson, A.D., 1992. External hyphae of vesicular – arbuscular mycorrhizal fungi associated with *Trifolium subterraneum* L. Spread of hyphae and phosphorus inflow into roots. *New Phytol.* **120**: 371 – 380.

Jalali, B.L. and Thareja, M.L. 1985. Plant growth responses to vesicular – arbuscular mycorrhizal inoculation in soils incorporated with rock phosphate. *Indian Phytopathol.* **38**: 306 – 310.

Janos, D.P. 1992. Heterogeneity and scale in tropical vesicular – arbuscular mycorrhiza formation. In: Read, D.J., Lewis, D.H., Fitter, A.H. and Alexander, I.J. (Eds.) *Mycorrhizas in ecosystems* .CAB International, Wallingford, U.K. pp. 276 – 282.

Jansa, J., Mozafar, A., Anken, T., Ruh, R., Sanders, I.R. and Frossard, E. 2002. Diversity and structure of AMF communities as affected by tillage in a temperate soil. *Mycorrhiza.* **12**: 225 – 234.

Jansa, J., Mozafar, A. and Frossard, E. 2003. Long distance transport of P and Zn through the hyphae of an arbuscular mycorhizal fungus in symbiosis with maize. *Agronomie* **23**: 481 – 488.

Jasper, D.A., Abbot, L.K. and Robson, A.D. 1989. Soil disturbance reduces the infectivity of external hyphae of vesicular – arbuscular mycorrhizal fungi. *New Phytol.* **112**: 93 – 99.

Jasper, D.A., Robson, A.D. and Abbott, L.K. 1979. Phosphorus and the formation of vesicular – arbuscular mycorrhizas. *Soil Biol. Biochem.* **11**: 501 – 505.

Jayakumari, T.R. and Potty, V.P. 2009. Influence of P fertilizer and mycorrhizal inoculation on the yield attributes of edible Canna (*Canna edulis* Ker). *J. Root Crops.* **35**: 112 – 115.

Jenson, A. and Jakobsen, I. 1980. The occurrence of vesicular – arbuscular mycorrhiza in barley and wheat grown in some Danish soils with different fertilizer treatments. *Plant Soil.* **55**: 403 – 414.

John, S., Sebastian, M., Alexander, D., Kurup, S.S. and Rajmohan, K.2007. Arbuscular mycorrhizal fungal associations in different species of sesame. In: Kesavachandran, R., Nazeem, P.A., Girija, D., John, P.S. and Peter, K.V. (Eds.) *Recent Trends in Horticultural Biotechnology.* New India Publishing Agency, New Delhi, India pp. 925 – 927.

Johnson, D., Ijdo Marleen, Genney, D.R., Anderson. I.C. and Alexander, I.J. 2005. How do plants regulate the function, community structure and diversity of mycorrhizal fungi. *J. Exptl. Bot.* **56**: 1751 – 1760.

Johnson, N.C. and Pfleger, F.L.1992. Vesicular – arbuscular mycorrhizae and cultural Stress In: Bethlenfalvay, G.B., Linderman, R.G. (Eds.) *Mycorrhizae in Sustainable Agriculture.* American Society of Agronomy, Madison Was pp. 1 – 27.

Joner, E.J. and Jakobsen, I. 1994. Contribution by two arbuscular mycorrhizal fungi to P uptake by cucumber (*Cucumis sativus* L.) from (32) P – labelled organic matter during mineralization in soil. *Plant Soil* **163**: 203 – 209.

Ju, X.T., Kou, C.L., Christie, P., Dou, Z.X. and Zhang, F.S. 2007. Changes in the soil environment from excessive application of fertilizers and manures to two contrasting intensive cropping system on the North China Plain. *Environ. Pollut.* **145**: 497 – 506.

Jung, G., Mungier, G., Diem Hoang, D. and Dommergues, Y.R. 1981. Polymer entrapped symbiotic microorganisms as inoculants for legumes and non-legumes (Abstr.) *Fifth North American Conference on Mycorrhizae* Quebec, Canada. p. 54.

Juniper, S. and Abbot, L.K. 1992. The effect of a change of soil salinity on growth of hyphae from spores of *Gigaspora decipiens* and *Scutellospora calospora* Abstracts. *International Symposium on Management of Mycorrhizas in Agriculture, Horticulture and Forestry*, University of Western Australia.

Kabir, Z., O'Halloran, I.P., Fyles, J.W. and Hamel, C. 1997. Seasonal changes of arbuscular mycorrhizal fungi as affected by tillage practices and fertilization: Hyphal density and mycorrhizal root colonization. *Plant Soil.* **192**: 285 – 293.

Kabir, Z., O'Halloran, I.P., Fyles, J.W. and Hamel, C. 1998. Dynamics of the mycorrhizal symbiosis of corn (*Zea mays* L.) : Effects of host physiology, tillage practice and fertilization on spatial distribution of extra-radical mycorrhizal hyphae in the field. *Agric. Ecosys. Environ.* **68**: 151 – 163.

Kahiluoto, H, Ketoja, E. and Vestberg, M. 2000. Promotion of utilization of arbuscular mycorrhiza through reduced P fertilization. 1. Bioassays in a growth chamber. *Plant Soil.* **227**: 191 – 206.

Kandasamy, D., Palanisamy, D. and Oblisami, G., 1988. Screening of germplasm of sweet potato for VA mycorrhizal fungal occurrence and response of the crop to inoculation of VAM fungi and *Azospirillum*. *J. Root Crops.* **14**: 37.

Karlen, D.L. and Whitney, D.A. 1980. Dry matter accumulation, mineral concentrations and nutrient distribution in winter wheat. *Agron J.* **72**: 281-288.

Karunasinghe, T.G., Fernando, W.C. and Jayasekera, L.R. 2009. The effect of poultry manure and inorganic fertilizer on arbuscular mycorrhiza in coconut. *J. Natn. Sci. Foundation Sri Lanka.* **37**: 277-279.

Kehri, H.K., Chandra, S. and Maheswari, S. 1987. Occurrence and intensity of VAM in weeds ornamentals and cultivated plants in Allahabad and areas adjoining it. In: Varma, A.K., Oka, A.K., Mukerji, K.G., Tilak, K.V.B.R. and Raj, J. (Eds.) Proc. *Natl. Workshop on Mycorrhiza* JNU, New Delhi. pp. 274 – 283.

Khalafallah, A.A. and Abo-Ghalia, H.H. 2008. Effect of arbuscular mycorrhizal fungi on the metabolic products and activity of antioxidant system in wheat plant subjected to short-term water stress followed by recovery at different growth stages. *J. Applied Sciences Res.* **4**: 559 – 569.

Khaliel, A.S. 1993. Influence of three *Glomus* species on growth and iron uptake of tomato seedlings. *Cyptogamic Botany.* **4**: 14-18.

Khaliel, A.S. and Elkhider, K.A. 1987. Response of tomato to inoculation with vesicular-arbuscular mycorrhiza. *Nordic J. Botany.* **7**: 215 – 218.

Khalil, S. and Loynachan, T.E. 1994. Soil drainage and distribution of VAM fungi in two toposequences. *Soil Biol Biochem.* **26**: 929 – 934.

Khalil, S., Loynachan, T.E. and Tabatabai, M.A.1994. Mycorrhizal dependency and nutrient uptake by improved and unimproved corn and soybean cultivars. *Agron. J.* **86**: 949 – 958.

Khaliq, A. and Sanders, F.E. 2000. Effects of vesicular – arbuscular mycorrhizal inoculation on the yield and phosphorus uptake of field grown barley. *Soil. Biol Biochem.* **32**: 1691 – 1696.

Khan, A.G. 1972. The effect of vesicular – arbuscular mycorrhizal association on growth of cereals I. Effects on maize growth. *New Phytol.* **71**: 613 – 619.

Khan, A.G. 1974. The occurrence of mycorrhiza in halophytes, hydrophytes and xerophytes and of *Endogone* spores in adjacent soil. *J. Gen. Microbiol.* **81**: 7 – 14.

Kim, C.K. and Weber, D.J. 1985. Distribution of VA mycorrhiza on halophytes on inland salt playas. *Plant Soil* **83**: 207 – 214.

Kinden, D.A. and Brown, M.F. 1975. Electron microscopy of vesicular arbuscular mycorrhizae of yellow poplar. III. Host – endophyte interactions during arbuscular development. *Can. J. Microbiol.* **21**: 1930 – 1939.

Klironomos, J.N., Hart, M.M. Gurney, J.E. and Moutoglis, P., 2001. Interspecific differences in the tolerance of arbuscular mycorrhizal fungi to freezing and drying. *Can J. Bot.* **79**: 1161 – 1166.

Kohel, R.J.1978. Survey of *Gossypium hirsutum* L germplasm collection for seed oil percentage and seed characteristics. *ARS Series* **187**: 38 p.

Koide, R.T. and Schreiner, R.P. 1992. Regulation of the vesicular – arbuscular mycorrhizal symbiosis. *Ann. Rev. Pl. Physiol. Mol. Biol.* **43**: 557 – 581.

Konde, B.K., Tambe, A.D. and Ruikar, S.K. 1988. Yield of nitrogen and phosphorus uptake by onion as influenced by inoculation of VAM fungi and *Azospirillum brasilense*. In: Mahadevan, A., Raman, N., Natarajan, K. (Eds.) *Mycorrhiza for Green Asia*. Proc. of the First Asian Conference on Mycorrhiza, University of Madras, Chennai, India 29 – 31 January 1988. pp. 222 – 224.

Koske, R.E. and Tessier, B. 1983. A convenient permanent slide mounting medium. *Mycol. Soc. Am. Newsl.* **34**: 59.

Kothari, S., Marschner, H. and Romheld, V.1991. Contribution of the VA mycorrhizal hyphae in acquisition of phosphorus and zinc by maize grown in a calcareous soil. *Plant Soil*. **131**: 177 – 185.

Krishna, K.R. and Bagyaraj, D.J. 1984. Growth and nutrient uptake of peanut inoculated with the mycorrhizal fungus *Glomus fasciculatum* compared with the non-mycorrhizal ones. *Plant Soil* **77**: 405 – 408.

Krishna, K.R., Shetty, K.G., Daft, P.J. and Andrews, D.J. 1985. Genotype dependent variation in mycorrhizal colonization and response to inoculation in pearl millet. *Plant Soil*. **86**: 113 – 125.

Kruckelmann, H.W. 1973. Die vesikular – arbuskulare mykorrhiza und ihre Beeinflussung in land wirtschaftlichen. Kulturen. Dissertation, Universität Braunschweig.

Kruckelmann, H.W. 1975. Effects of fertilizers, soils, soil tillage and plant species on the frequency of *Endogone* chlamydospores and mycorrhizal infection in arable soils. In: Sanders, F.E., Mosse, B. and Tinker, P.B. (Eds.) *Endomycorrhizas*. Academic Press, London. pp. 511 – 515.

Kumaresan, S. 1997. Studies on VA mycorrhizal association in mulberry in different ecosystems of Kerala State India. Ph.D. thesis, Bharathidasan University, Tiruchirapalli, India. p. 88.

Kumutha, K., Kandasamy, D. and Rangarajan, M. 1993. Influence of VA mycorrhizae and *Azospirillum brasilense* on mulberry. *Indian Seri. Culture*. **32**: 238 – 241.

Land, S., Schöönbeck, F. 1991. Influence of different soil types on abundance and seasonal dynamics of vesicular arbuscular mycorrhizal fungi in arable soils of North Germany. *Mycorrhiza* **1:** 39 – 44.

Lemcke – Norojarvi, M., Kamal – Eldin, A., Appelqvist, I.A., Dimberg, L.H., Ohrvall, M. and Vessby, B. 2001. Corn and sesame oils increase serum gamma – to copherol concentrations in healthy Swedish women. *J. Nutr.* **131:** 1195 – 1201.

Levy, Y. and Krikun, J. 2006. Effect of vesicular arbuscular mycorrhiza on *Citrus jambhiri* water relations. *New Photol.* **85:** 25 – 31.

Leye, M. 2006. Résponse du sésame (*Sesamum indicum* L.) á Í inoculation mycorhizienne arbusculaire. Memoire de DEA, Université Cheikh anta Diop de Darkar, Sénégal, p. 78.

Li, X -L. and Marschner, H., George, E. 1991. Phosphorus depletion and pH decrease at the root-soil and hyphae soil interfaces of VA mycorrhizal while clover fertilized with ammonium. *New Phytol.* **119:** 397 – 404.

Li-L-F., Zhang, Y. and Zhao, Z-W. 2007. Arbuscular mycorrhizal colonization and spore density across different land-use types in a hot and arid ecosystem, Southwest China. *J. Pl. Nut. Soil Sci.* **170:** 419-425.

Lindermann, R.C. and Hendix, J.W. 1982. Evaluation of plant response to colonization by vesicular – arbuscular mycorrhizal fungi. In: Schenck, N.C. (Ed.) *Methods and Principles of Mycorrhizal Research.* American Phytopathological Society St Paul, Minnesota.

Lioussanne, L., Jolicoeur, M. and St. Arnaud, M. 2008. Mycorrhizal colonization with *Glomus intraradices* and development stage of transformed tomato roots significantly modify the chemotactic response of zoospores of the pathogen *Phytophthora nicotiane*. *Soil Biol. Biochem.* **40:** 2217 – 2224.

Liu, Y.L,, Li, Y., Matsubara, Y., Inayaki, M. and Sugiyama, M. 2006. Promotion of rooting and the increase in leaf GABA content of mycorrhizal tea plant. *Acta Hort. ISHS* **761:** 267 – 270.

Logi, C., Sbrana, C. and Giovannetti, M. 1998. Cellular events involved in survival of individual arbuscular mycorrhizal symbionts growing in the absence of the host. *Appl. Environ. Microbiol.* **64**: 3473 – 3479.

López-Sánchez, M.E. and Honrubia, M. 1992. Seasonal variation of vesicular arbuscular mycorrhizae in eroded soil from southern Spain. Mycorrhiza. **2**: 33-39.

Lovelock, C.E. and Miller, R. 2002. Heterogeneity in inoculum potential and effectiveness of arbuscular mycorrhizal fungi. *Ecology.* **83**: 823 – 832.

Lu, S. and Miller, M.H. 1989. The role of VA mycorrhizae in the absorption of P and Zn by maize in field and growth chamber experiments *Can. J. Soil. Sci.* **69**: 97 – 109.

Mäder, P., Edenhofer, S., Boller, T., Wiemken, A. and Niggli, U. 2000. Arbuscular mycorrhizae in a long- term field trial comparing low-input (organic, biological) and high-input (conventional) farming systems in a crop rotation. *Biol. Fertil. Soils.* **31**: 150 – 156.

Malibari, A.A., Al-Fassi, F.A. and Ramadan, E.M. 1988. Incidence and infectivity of vesicular arbuscular mycorrhiza in some Saudi soils. *Plant Soil.* **112**: 105-111.

Manoharachary, C. and Prakash, P. 1991. Vesicular – arbuscular mycorrhizal symbiosis and its role in nutrition of oil seed crops of semi arid tropics. *J. Soil Biol. Ecol.* **11**: 84 – 89.

Marschner, H. and Dell, B. 1994. Nutrient uptake in mycorrhizal symbiosis. *Plant Soil.* **159**: 89 – 102.

Mathur, N. and Vyas, A. 1996. Changes in free amino acid levels in *Prosopis cineraria* by VA mycorrhiza. *Natl. Acad. Sci. Lett.* **19**: 1 – 3.

Matsubara, Yoh-ichi, Ishgaki, T. and Koshikawa, K. 2008. Changes in free amino acid concentration in mycorrhizal strawberry plants. *Sci. Hort.* **119**: 392-396.

McGonigle, T.P. and Miller, M.H. 1993. Mycorrhizal development and phosphorus absorption in maize under conventional and reduced tillage. *Soil. Sci. Soc. Am. J.* **57**: 1002 – 1006.

McGonigle, T.P. and Miller, M.H. 1999. Winter survival of extradical hyphae and spores of arbuscular mycorrhizal fungi in the field. *Appl. Soil Ecol.* **12**: 41 – 50.

McGonigle, T.P., Evans, D.G. and Miller, M.H. 1990. Effect of degree of soil disturbance on mycorrhizal colonization and phosphorus absorption by maize in growth chamber and field experiments. *New Phytol.* **116**: 629 – 636.

McGonigle, T.P., Yano, K. and Shinhama, T. 2003. Mycorrhizal phosphorus enhancement of plants in undistributed soil differs from phosphorus uptake stimulation by arbuscular mycorrhizae over non-mycorrhizal controls. *Biol. Fertil. Soils* **37**: 268-273.

McGorigle, T.P. and Fitter, A.H. 1990. Ecological specificity of vesicular – arbuscular mycorrhizal associations. *Mycol. Res.* **94**: 120 – 122.

McGraw, A.C. and Schenck, N.C. 1981. The mycorrhizal relationship in tomato as influenced by fungal species, inoculum density and cultivar. In: Fortin, J.A (Ed.) *Proc. of the Fifth North American Conference on Myconhizae.* 16 – 21 Aug University of Laval, Quebec, Canada. pp. 19.

McMillen, B.G., Juniper, S. and Abbot, L.K. 1998. Inhibition of hyphal growth of a vesicular – arbuscular mycorrhizal fungus in soil containing sodium chloride limits the spread of infection from spores. *Soil Biol. Biochem.* **30**: 1639 – 1646.

Mehraveran, H., 1977. Mycorrhizal dependency of six citrus cultivars. Ph.D. Thesis, University of Illinios, U.S.A.

Mejstrick, V.K. 1972. Vesicular – arbuscular mycorrhizas of the species of a *Molinietum coeruleae* L. I. Association. the ecology. *New Phytol.* **71**: 883 – 890.

Meloh, K.A. 1961. Untersuchungen zur Biologie und Bedeutung der endotrophen Mycorrhiza bei *Zea mays* L. und *Avena sativa* L. Dissertationsschrift der Universtität Köln. Meloh, K.A. 1963. *Untersuchungen zur Biologie der endotrophen Mycorrhiza bei Zea mays* L. und *Avena sativa* L. *Arch. Microbiol.* **46**: 369 – 381.

Menge, J.A. 1983. Utilization of vesicular – arbuscular mycorrhizal fungi in agriculture. *Can. J. Bot.* **61**: 1015 – 1024.

Menge, J.A. 1984. Inoculum production. In: Powell, C.D. and Bagyaraj, D.J. (Eds.) *VA Mycorrhiza*. CRC Press Inc. Boca Raton, Florida USA. pp. 188 – 199.

Menge, J.A., Johnson, E.L.V. and Platt, R.G. 1978. Mycorrhizal dependency of several citrus cultivars under three nutrient regimes. *New Phytol.* **81**: 553 – 559.

Mercy, M.A., Shivashankar, G. and Bagyaraj, D.J. 1990. Mycorrhizal colonization in cowpea is host dependent and heritable. *Plant Soil.* **121**: 292 – 294.

Miyahara, Y., Hibasami, H., Katsuzaki, H., Imai, K. and Komiya, T. 2001. Sesamolin from sesame seed inhibits proliferation by inducing apoptosis in human lymphoid leukemia Molt 4B cells. *Int. J. Mol. Med.* **7**: 369 – 371.

Mohammad, M.J., Pan, W.L. and Kennedy, A.C. 1998. Seasonal mycorrhizal colonization of winter wheat and its effect on wheat growth under dry land field conditions. *Mycorrhiza*. **8**: 139 – 144.

Mohankumar, V. and Mahadevan, A. 1986. Survey of vesicular arbuscular mycorrhizae in mangrove vegetation. *Curr. Sci.* **55**: 936.

Mohankumar, V. and Mahadevan, A. 1988. Seasonal changes in spore density and root colonization of VAM in a tropical forest In: Mahadevan, A., Raman, N. and Natarajan, K (Eds.) Mycorrhizae *for Green Asia* I ACOM, Madras, India. pp. 80 – 81.

Mohankumar, V. and Mahadevan, A. 1988. Viability of VAM spores in a tropical forest soils. In: Mahadevan, A., Raman, N. and Natarajan, K (Eds.) Mycorrhizae *for Green Asia* I ACOM, Madras, India. pp. 80 – 81.

Morone – Fortunato, I. and Avato, P. 2008. Plant development and synthesis of essential oils in micropropagated and mycorrhiza inoculated plants of *Origanum vulgare* L. ssp. *hirtum* (Link)/etswaart. *Plant, Cell, Tissue and Organ Culture.* **93**: 139 – 149.

Morton, J.B. and Benny, G.L. 1990. Revised classification of arbuscular mycorrhizal fungi (zygomycetes) a new order, Glomales, two new suborders, Glomineae and Gigasporineae and two new families, Acaulosporaceae and Gigasporaceae, with an emendation of Glomaceae. *Mycotaxon.* **37**: 471-491.

Morton, J.B., Bentivenga, S.P. and Wheeler, W.W. 1993. Germplasm in the international collection of arbuscular and vesicular-arbuscular mycorrhizal fungi (INVAM) and procedures for culture development, documentation and storage. *Mycotaxon* **48**: 491 – 528.

Morton, J.B. and Redecker, D. 2002. Two new families of Glomales, Archaeosporaceae and Paraglomaceae with two new genera *Archaeospora* and *Paraglomus* based on concordant molecular and morphological characters, *Mycologia* **93**: 181-195.

Mosse, B. 1957. Growth and chemical composition of mycorrhizal and non-mycorrhizal apples. *Nature.* **179**: 922.

Mosse, B. 1972. Effects of different *Endogone* strains on the growth of *Paspalum notatum. Nature.* **239**: 221 – 223.

Mosse, B. 1981. Vesicular - arbuscular mycorrhiza research for tropical agriculture.Research Bulletin No. 194, Hawaii Institute Tropical Agriculture and Human Resources. pp. 84.

Mosse, B. and Hayman, D.S. 1980. Mycorrhiza in agricultural plants. In: Mikola, P (Ed.), *Tropical Mycorrhiza Research*, Clarendon Press. pp 213 – 230.

Mosse, B.1973. Advances in the study of vesicular arbuscular mycorrhiza. *Ann. Rev. Phytopathol.* **11**: 171 – 196.

Mosse, B.1973. Plant growth response to vesicular – arbuscular mycorrhiza IV. In soil given additional phosphate. *New Phytol.* **72**: 127 – 136.

Muthukumar, T. and Udaiyan, K. 2000. Growth and yield of cowpea as influenced by changes in arbuscular mycorrhiza in response to organic manuring. *J. Agron. Crop Sci.* **188**: 123 – 132.

Nayar, N.M. 1984. Sesame. In: Simmonds NW (Ed.) *Evolution of Crop Plants.* Longman London. pp. 231 – 233.

Neeraj and Singh, K. 2008. Biochemical changes in *Phaseolus vulgaris* L dual inoculated with AM fungi and *Rhizobium. Indian J. Bot. Res.* **4**: 73 – 80.

Negi, M., Sachdev, M.S. and Tilak, K.V.B.R. 1990. Influence of soluble phosphorus fertilizer on the interaction between vesicular – arbuscular mycorrhizal fungus and *Azospirillum brasilense* in barley (*Hordeum vulgare* L). *Biol. Fertil. Soils.* **10**: 57.

Nelsen, C.E. and Safir, G.R., 1982. Increased drought tolerance of mycorrhizal onion plants caused by improved phosphorus nutrition. *Planta* 154: 407 – 413.

Nielsen, J.D. and Jensen, A. 1983. Influence of vesicular – arbuscular mycorrhiza fungi on growth and uptake of various nutrients as well as upake ratio of fertilizer P for lucerne *(Medicago sativum). Plant Soil.***70**: 165 – 172.

Novoa, R. and Loomis, R.S. 1981. Nitrogen and plant production. *Plant Soil* **58**: 177 – 204.

Oliver, A.J., Smith, S.E., Nicholas, D.J.D., Wallace, W. and Smith, F.A. 1983. Activity of nitrate reductase in *Trifolium subterraneum* – Effect of mycorrhizal infection and phosphate nutrition. *New Phytol.* **94:** 63 – 79.

Pacovsky, R.S. 1989. Carbohydrate, protein and amino acids status of Glycine – *Glomus – Bradyrhizobium* symbiosis. *Physiol. Plant.* **75**: 346 – 354.

Panja, B. and Chaudhuri, S. 1999. Interaction between organic manures and arbuscular mycorrhiza in *Cajanus* root association in alluvial soil In: *Proc. Natl. Conf. Mycorrhiza,* Bhopal, India 5 – 7 March, 1999. p. 50.

Pattinson, G.S. and McGee, P.A.1997. High densities of arbuscular mycorrhizal fungi maintained during long fallows in soils used to grow cotton except when soil is wetted periodically. *New Phytol.* **136**: 571 – 580.

Peuss, H. 1958. Untersuchungen zur Ökologie und Bedeutung der Tabakmycorrhiza. *Arch Microbiol.* **29**: 112 – 142.

Phillips, J.M. and Hayman, D.S. 1970. Improved procedure for clearing roots and staining parasitic and vesicular arbuscular mycorrhizal fungi for rapid assessment of infection. *Trans. Br. Mycol. Soc.* **55**: 158 – 161.

Piper, C.S. 1942. *Soil and Plant Analyses*, University of Adelaide.

Plenchette, C., Fortin, J.A. and Furlan, V. 1983. Growth response of several plants species to mycorrhizae in a soil of moderate fertility. *Plant Soil.* **70**: 199 – 209.

Plenchette, C., Furlan, V. and Fortin, J.A. 1983. Responses of endomycorrhizal plants grown in a calcined montmorillonite clay to different levels of soluble phosphorus I. Effect on growth and mycorrhizal development. *Can J. Bot.* **61** : 1377 – 1383.

Pond, E.C., Menge, J.A. and Jarrell, W.M. 1984. Improved growth of tomato in salinized soil by vesicular – arbuscular mycorrhizal fungi collected from saline soils. *Mycologia* **76**: 74 – 84.

Porter, W.M., Abbot, L.K. and Robson, A.D. 1978. Field Survey of the distribution of VA mycorrhizal fungi in relation to soil pH. *J. Appl. Ecol.* **24**: 659 – 662.

Potty, V.P. 1988. Response of Cassava (*Manihot esculenta*) to VAM inoculation in acid laterite soil. *In:* Mahadevan, A., Raman, N. and Natarajan, K. (Eds.) *Mycorrhizae for Green Asia.* CAS in Botany, University of Madras, Madras pp. 246 – 249.

Potty, V.P. 1990b. VA mycorrhizal association in tuber crops and their role in crop production. In: *Annu. Progress Rep.* CTCRI, Sreekariyam, Thiruvananthapuram. pp. 81 – 83.

Potty, V.P. and Harikumar. V.S. 1995. Interaction of vesicular – arbuscular mycorrhizal fungi and phosphobacterium in sweet potato rhizosphere. In: Adholeya, A. and Singh, S (Eds.) *Mycorrhizae: biofertilizers for the future.* Proceedings of the Third National Conference on Mycorrhiza, Tata Energy Research Institute, New Delhi, 13 – 14 March 1995. pp. 177 – 179.

Potty, V.P.1978. Occurrence of vesicular – arbuscular mycorrhiza in certain tuber crops. *J. Root Crops.* **4**: 49 – 50.

Prakash, A. and Tandon, V. 2002. Exploiting mycorrhiza for oil seed production. Biotechnology of microbes and sustainable utilization. Scientific Publishers (India), Jodhpur.

Prakash, A., Tandon, V. and Sharma, N.C. 2004. Effect of rock phosphate and VAM inoculation on growth and nutrient uptake in *Sesamum indicum* L. *Physiol. Mol. Biol. Plants*. **10**: 137 – 141.

Pravankumar, P., Shailaja, K.M., Rao, M.S. and Reddy, R.S. 1998. Genotype dependent variation in VAM infection and growth response of twelve cultivars of sesame *(Sesamum indicum L.). J. Indian Bot. Soc.* **77**: 71 – 74.

Peyronel, B. 1924. *Prime ricerche sulle micorrize endotrofiche sulla microflora radicicola normale delle fancrogame. Riv. Biol.* **6**: 1

Rabatin, S.C. 1979. Seasonal and edaphic variation in vesicular- arbuscular mycorrhizal infection of grasses by *Glomus tenuis*. *New Phytol*. **83:** 95-102.

Rakshit, A. and Bhadoria, P.B.S. 2008. Indigenous arbuscular mycorrhiza is more important for early growth period of groundnut (*Arachis hypogea* L.) for influx in an Oxisol. *Acta Agric. Slov.* **91**: 397 – 406.

Ramana, B.V. and Babu, R.S.H., 1999. Response of onion (*Allium cepa*) to inoculation of VAM fungi at different levels of phosphorus In*: Proceedings of the National Conference on Mycorrhiza.* 5 – 7 March Bhopal, India.

Rao, A.S. and Parvathi, K. 1982. Development of VA mycorrhiza in groundnut and other hosts. *Plant Soil* **66**: 133 – 137.

Rao, N.S.S., Tilak, K.V.B.R. and Singh, C.S. 1985. Effect of combined inoculation of *Azospirillum brasilense* and vesicular-arbuscular mycorrhiza on pearl millet (*Pennisetum americanum*) *Plant Soil* **84**: 283 – 286.

Ratti, N. and Janardhanan, K.K. 1996. Response of dual inoculation with VAM and *Azospirillum* on the yield and oil content of palmarosa (*Cymbopogon matrinni* var. *motia*). *Microbiol. Res.* **151**: 325 – 328.

Ratti, N., Kumar, S., Verma, H.N. and Gautam, S.P. 2001. Improvement of bioavailability of tricalcium phosphate to *Cymbopogon martinni* var. *motia* by rhizobacteria AMF and *Azospirillum* inoculation. *Microbiol. Res.* **156**: 145 – 149.

Reddy, C.N., Bharathi, B.K., Rajkumar, H.G. and Sunanda, D.N. 2004. Infectivity efficacy of four native vesicular – arbuscular mycorrhiza fungi on sugarcane. *Mycorrhiza News* **16**: 9 – 12.

Reddy, S.R., Rachel, E.K. and Reddy, S.M. 1997. Effect of water stress on VAM colonization and growth of sunflower. *J. Mycol. Pl. Pathol.* **27**: 294-296.

Reid, C.P.P., Bowen, G.D. 1979. Effects of soil moisture on VA mycorrhizal formation and root development in *Medicago*. In: Harley, J.C.and Russel, R.S. (eds.) *The Soil Root Interface*. Academic Press, London. pp. 211 – 219.

Robson, A.D. and Abbot, L.K.1989. The effect of soil acidity on mycorrhizal activity in soil In: Robson, A.D. (Ed.) *Soil Acidity and Plant Growth*. Academic Press, Sydney. pp. 139 – 165.

Rohaydi, A, Smith, F.A., Murray, RS. and Smith, S.E. 2004. Effects of pH on mycorrhizal colonization and nutrients uptake in cowpea under conditions that minimize confounding effects of elevated available aluminium. *Plant Soil*. **260**: 283-290.

Rolin, D., Pfeffer, P.E., Douds, D.D., Farrel, H.M. and Shachar, H.Y. 2001. Arbuscular mycorrhizal symbiosis and phosphorus nutrition. Effects on amino acid production and turnover in leek. *Symbiosis*. **30**: 1 – 14.

Römer, W. and Schilling, G. 1986. Phosphorus requirement of the wheat plant in various stages of its life cycle. *Plant Soil* **91**: 221 – 229.

Rozema, J., Arp, W. Van Diggelen, J., Van Esbroek, M., Broekman, R. and Punte, H. 1986. Occurrence and ecological significance of vesicular – arbuscular mycorrhiza in the salt marsh environment. *Acta Bot. Neerl.* **35**: 457 – 467.

Ruiz – Lozano, J.M. and Azcón .R., 1996. Viability and infectivity of mycorrhizal spores after long-term storage in soils with different water potentials. *Applied Soil Ecol.* **3**: 183 – 186.

Russel, E.J. 2007. *Soil Condition and Plant Growth*. Biotech Books, New Delhi.

Russo, A., Felici, C., Toffanin, A., Götz, M., Collados, C., Barea, J.M., Moënne – Loccoz, Y., Smalla, K., Vanderbyden, J. and Nuti, M. 2005. Effect of *Azospirillum* inoculants on arbuscular mycorrhiza establishment in wheat and maize plants. *Biol. Fertil. Soils.* **41**: 301 – 309.

Same, B.I., Robson, A.D. and Abbot, L.K. 1983. Phosphorus soluble carbohydrate and endomycorrhizal infection. *Soil Biol. Biochem.* **15**: 593 – 597.

Sanders, F.E. 1975. The effect of foliar – applied phosphate on the mycorrhizal infections of onion roots. In: Sanders, F.E, Mosse, B. and Tinker, P.B. (Eds.) *Endomycorrhizas*. Academic Press, London, U.K. pp. 261 – 276.

Sanders, F.E. and Tinker, P.B. 1971. Mechanism of absorption of phosphate from soil by *Endogone* mycorrhizas. *Nature.* **233**: 278 – 279.

Sanders, F.E., Tinker, P.B., Black, R.L.B. and Palmerby, S.M. 1977. The development of endomycorrhizal root systems. I. Spread of infection and growth promoting effects with four species of vesicular – arbuscular endophytes. *New Phytol.* **78**: 257 – 268.

Sanders, I.R. and Fitter, A.H. 1992. The ecology and functioning of vesicular-arbuscular mycorrhizas in co-existing grassland species II. Nutrient uptake and growth of vesicular- arbuscular mycorrhizal plants in semi- natural grassland. *New Phytol.* **120**: 525-533.

Sankar, D., Sambandam, G., Rao, M.R. and Pugalendi, K.V. 2004. Impact of sesame oil on nifedipine in modulating oxidative stress and electrolytes in hypertensive patients. *Asia Pac. J. Clin. Nutr.* **13**: 107.

Schenck, N.C. and Pérez, Y. 1987. Manual for the identification of VA mycorrhizal fungi INVAM, Florida University, Gainesville. F.L., U.S.A.

Schenck, N.C., Kinloch, R.A. and Dickson, D.W. 1975. Interaction of endomycorrhizal fungi and root-knot nematode on soybean In: Sanders, F.E, Mosse, B. Tinker, P.B. (Eds.) *Endomycorrhizas*. Academic Press, New York. pp. 607 – 617.

Schreiner, R.P. 2007. Effect of nature and non-native arbuscular mycorrhizal fungi on growth and nutrient uptake by Pinot root (*Vitis vinefera* L) in two soils with contrasting levels of phosphorus. *Appl. Soil Ecol.* **36**: 205 – 215.

Schreiner, R.P. and Linderman, R.G. 2005. Mycorrhizal colonization in dryland vineyards of the Willamette valley, Oregon. *Small Fruit Rev.* **4**: 41 – 55.

Schubert, A., Wyss, P. and Wiemken, A. 1992. Occurrence of trehalose in vesicular –arbuscular mycorrhizal fungi and in mycorrhizal roots. *J. Plant Physiol.* **140**:41 – 45.

Schüßler, A., Schwarzott, D. and Walker, C. 2001. A new fungal phylum, the Glomeromycota: phylogeny and evolution. *Mycol. Res.* **2001**: 1413 – 1421.

Schwab, S.M., Menge, J.A. and Leonard, R.T. 1983. Comparison of stages of vesicular -arbuscular mycorrhizaa formation in sudangrass grown at two levels of phosphorus nutrition. *American J. Bot.* **70**: 1225 – 1232.

Schwab, S.M., Menge, J.A. and Leonard, R.T. 1983. Quantitative and qualitative effects of phosphorus on extracts and exudates of sudangrass roots in relation to vesicular – arbuscular mycorrhiza formation. *Plant Physiol.* **73**: 761 – 765.

Secilica, J. and Bagyaraj, D.J. 1994. Evaluation and first – year field testing of efficient vesicular – arbuscular mycorrhizal fungi for inoculation of wetland rice seedlings. *World J. Microbiol. Biotechnol.* **10**: 381 – 384.

Selvaraj, T. and Subramanian, G. 1988. Light and scanning electron microscopic studies of VAM in *Sesamum indicum* roots *In:* Mahadevan, A., Raman, N. and Natarajan, K. (Eds.) *Mycorrhizae for Green Asia*. I ACOM, Madras, India. pp.106 – 110.

Selvaraj, T. and Subramanian, G. 1990. Phenols and lipids in mycorrhizal and non mycorrhizal roots of *Sesamum indicum*. *Curr Sci.* **59**: 471 – 473.

Selvaraj, T. and Subramanian, G. 1995. Autofluorescence of vesicular-arbuscular mycorrhizal fungi in roots and rhizosphere soil of *Acrous calamus* and *Sesamum indicum*. *Acta Bot. Indica.* **23**: 87 – 91.

Sengupta A. and Chaudhuri, S. 1990. Vesicular arbuscular mycorrhiza (VAM) in pioneer salt marsh plants of Ganges river delta in West Bengal (India). *Plant Soil*. 122: 111-113.

Shokri, S. and Maadi, B. 2009. Effects of arbuscular mycorrhizal fungus on the mineral nutrition and yield of *Trifolium alexandrinum* plants under salinity stress. *J. Agron*. **8**: 79 – 83.

Sieverding, E. and Syverston, S. 1983. Influence of soil regimes on VAM II: Effects of soil temperature and water regime on growth, nutrient uptake and water utilization of *Eupatorium odoratum* Z. *Acker – Pflanzenbaw*. **152**: 56 – 67.

Singh, A.P., Chathurvedi, S., Tripathi, M.K. and Singh, S. 2004. Growth and yield of greengram (*Vigna radiata* (L.) Wilczek) as influenced by biofertilizer and phosphorus application. *Ann Biol*. **20**: 227-229.

Singh, C.S. 1996. Arbuscular mycorrhiza (AM) in association with *Rhizobium* sp improves inoculation, N_2 fixation and N utilization of pigeon pea (*Cajanus cajan*) as assessed with a15 N technique in pots. *Microbiol. Res*. **151**: 87-92.

Siqueira, J.O., Saggin – Júnior, O.J., Flores Aylas, W.W. and Guimarães, T.G.1998. Arbuscular mycorrhizal inoculation and super phosphate application influence plant development and yield of coffee in Brazil. *Mycorrhiza*. **7**: 293 – 300.

Siqueria, J.O. and Paula, M.A. 1986. Effito de micorrizas vesiculo – arbusculares na Nutrica e aproveitamento de josforo pela soja em solo sob cenaddo. R. bras. Ci Solo. **10**: 97 – 102. Robert.1985. *Umweltforschung zu Waldsba den Et De Bundesminister fur Forschung and Technologic offent lichkeitas – arbeit*. New Press Verlag Glvellsehaft mbH, Passau, Bonn, pp. 1 – 79.

Sivaprasad, P., Alice, A., Inasi, K.A. and Kunju, U.M. 1990. Vesicular – arbuscular mycorrhizal colonization of tuber crops grown in the reclaimed *Kayal* land. *J. Root Crops* ISRC Nat. Sym. Special. **17**: 134 – 136.

Sivaprasad, P., Sulochana, K.K. and Nair, S.K. 1990. Comparative efficiency of differentVA mycorrhizal fungi on cassava (*Manihot esculenta* Crantz.) *J. Root Crops* 16: 39 – 40.

Smith, K.P., Goodman, R.M. 1999. Host variation for interactions with beneficial Plant associated microbes. *Annul. Rev. Phytopathol.* **37**: 473 – 491.

Smith, S.E. 1982. Inflow of phosphate into mycorrhizal and non-mycorrhizal plants of *Trifolium subterraneum* at different levels of soil phosphate. *New Phytol.* **90**: 293 – 303.

Smith, S.E. and Gianinazzi – Pearson, V. 1988. Physiological interactions between symbionts in vesicular – arbuscular mycorrhizal plants. *Ann. Rev. Pl. Physiol. Mol. Biol.* **39**: 221 – 244.

Smith, S.E. and Read, D.J. 1997. *Mycorrhizal Symbiosis*, 2 edn, Academic Press, London.

Smith, S.E. and Walker, N.E. 1981. A qualitative study of mycorrhizal infection in *Trifolium*. Separate determination of the rates of infection and mycelial growth. *New Phytol.* **89**: 225 – 240.

Smith, S.E., Smith, F.A. and Jakobsen, I. 2003. Mycorrhizal fungi can dominate phosphate supply to plants irrespective of growth responses. *Plant Physiol.* **133**: 16-20.

Smith, S.E., Smith, F.A. and Jakobsen, I. 2004. Functional diversity in arbuscular mycorrhizal (AM) symbioses: the contribution of the mycorrhizal P uptake pathway is not correlated with mycorrhizal responses on growth or total P uptake. *New Phytol.* **162**: 511-524.

Snell, F.D. and Snell, C.T. 1949. *Colorimetric methods of analysis* Vol. II *Inorganic*. D. Van Nostrand Co. Inc. Princeton, New Jersey. pp. 813 – 815.

Soon, Y.K. and Arshad, M.A. 1996. Effect of cropping system on nitrogen, phosphorus and potassium forms and organic carbon in a Gray Luvisol. *Biol. Fertil. Soils.* **22**: 184 – 190.

Sorial, M.E. 2001. Growth, phosphorus uptake and water relations of wheat infected with an arbuscular mycorrhizal fungus under water stress. *Ann. of Agric Sci. Moshtohor.* **39**: 909 – 931.

Srinivas, K., Ramaraj, B. and Shanmugam, N. 1988. Survey for the occurrence of native VAM fungi. In: Mahadevan, A., Raman, N. and Natarajan, K (eds.) *Mycorrhizae for Green Asia,* CAS in Botany, University of Madras, Madras. pp. 111 – 113.

Steffens, G.L., Yang, S.Y., Steffens, C.L. and Brenna, T. 1963. Influence of paclobutrazol (PP 333) on apple seedling growth and physiology. *Proc. of the Plant Growth Regulation Society of America* **10**: 195 – 205.

Stewart, L.I., Hamel, C., Hogue, R. and Moutoglis, P. 2005. Response of strawberry to inoculation with arbuscular mycorrhizal fungi under very high soil phosphorus conditions. *Mycorrhiza* **15**: 612 – 619.

Strullu, D.G. and Plenchette, C. 1991. The entrapment of *Glomus* sp. in alginate beads and their use as root inoculum. *Mycol. Res.* **95**: 1194 – 1196.

Strzemska, J. 1975. Mycorrhiza in farm crop grown in monoculture. In: Sanders, F.E, Mosse, B., Tinker, P.B. (Eds.) *Endomycorrhizas.* Academic Press, London. pp. 527 – 535.

Subbaiah, B.V. and Asija, G.L. 1956. A rapid procedure for estimation of available nitrogen in soils. *Curr. Sci.* **25**: 255 – 260.

Stanford, S. and English, S. 1949. Use of flame photometer in rapid soil tests of K and Ca. *Agronomy J.* **141**: 446 – 497.

Subhashini, D.V., Rana, B.S. and Potty, V.P. 1988. Genotype dependent variation in mycorrhizal colonization and response to phosphorus in *Sorghum bicolor* In: Mahadevan, A., Raman, N. and Natarajan, K (Eds.) *Mycorrhizae for Green Asia* CAS in Botany, University of Madras, Madras. pp. 255 – 259.

Subramanian, K.S. and Charest, C. 1999. Acquisition of N by external hyphae of an arbuscular mycorrhizal fungus and its impact on physiological responses in maize under drought -stressed and well-watered conditions. *Mycorrhiza.* **9**: 69 – 75.

Suja, K.P., Abraham, J.T., Thamizh, S.N., Jayalekshmi, A. and Arumughan, C. 2004. Antioxidant efficacy of sesame cake extract in vegetable oil protection. *Food Chem.* **84**: 393 – 400.

Sulochana, T. and Manoharachary, C. 1989. VAM associations of castor and safflower. *Curr. Sci.* **58**: 449 – 461.

Sulochana, T., Reddy, P., Jagan Mohan, Manoharachary, C. 2000. Impact of season on the dynamics of arbuscular mycorrhizal fungi in sesamum. *J. Indian Bot. Soc.* **79**: 179 – 183.

Sundaresan, P., Raja, N.V., Gunasekaran, P. and Lakshmanan, M. 1987. Mycorrhizal association in Cow pea (*Vigna unguiculata*) increase phosphate uptake under water stress condition. In: Varma, A.K., Oka, A.K., Mukerji, K.G., Tilak, K.V.B.R and Janak, R. (Eds.) *Mycorrhiza Round Table* Proceedings of National Workshop, IDRC. Canada. pp. 461 – 470.

Sureshkumar, K.V., Harikumar, V.S. and Gopalakrishnan, P.K. 1995. Influence of host variety and edaphic factors on vesicular – arbuscular mycorrhizal (VAM) association in pigeonpea. *Acta Bot. Indica.* **23**: 81 – 85.

Sutton, J.C. 1973. Development of vesicular arbuscular mycorrhizae in crop plant. *Can J. Bot.* **51**: 2487-2493.

Suvercha and Mukerji, K.G. 1988. Influence of VAM fungi on the growth of two cultivars of *Capsicum annum*. In: Mahadevan, A., Raman, N. and Natarajan, K. (Eds.) *Mycorrhizae for Green Asia*. CAS in Botany, University of Madras, Madras. pp. 262 – 264.

Sylvia, D.M. and Williams, S.E.1992. Vesicular – arbuscular mycorrhizae and environmental stress. In: Bethlenfalvay, G.J. and Lindermann, R.G. (Eds.) *Mycorrhizae in Sustainable Agriculture*. ASA special publication, Madison. pp. 101 – 124.

Sylvia, D.M., Hammond, L.C., Bennett, J.M., Hass, J.H. and Linda, S.B.1993. Field response of maize to a VAM fungus and water management. *Agron J.* **85**: 193 – 198.

Tandon, H.L.S.1987. Phosphorus research and agricultural production in India, FDCO, New Delhi.

Tandon, V. and Prakash, A. 1998. Influence of soil inoculation with VA mycorrhiza and phosphorus solubilizing microorganisms on growth and phosphorus uptake in *Sesamum indicum*. *Int. J. Trop. Agric.* **16**: 201 – 209.

Tang, Z.Y. and Chen, A.J. 1986. Effect of mycorrhizal inoculation on insoluble phosphate absorption by *Citrus* seedlings on red earth. *Acta. Hort. Scinicia.* **13**: 75 – 79.

Tawaraya, K., Wagatsuma, T. and Sasai, K. 1990. Significance of aminoacids both in rhizosphere and in cytoplasm of root for vesicular arbuscular mycorrhizal infection. *Eighth North American Conference on Mycorrhiza Innovation and Hierarchial Integration, Jackson, Wyoming U.S.A., 5 – 8 September 1990.* University of Wyoming p. 278.

Tewari, L., Johri, B.N. and Tandon, S.M. 1993. Host genotype dependency and growth enhancing ability of VA – mycorrhizal fungi for *Eleusine coracana* (finger millet). *World J. Microbiol. Biotechnol.* **9**: 191 – 195.

Thaxter, R. 1922. A revision of the Endogonaceae. *Proc. Amer. Acad. Arts Sci.* **57**: 291-351.

Thiagarajan, T.R. and Ahmad, M.H., 1994. Phosphatase activity and cytokinin content in cowpeas (*Vigna unguiculata*) inoculated with a vesicular arbuscular mycorrhizal fungus. *Biol. Fertil. Soils.* **17**: 51 – 56.

Thomas, G.V. and Ghai, S.K. 1987. Genotype dependent variation in vesicular – arbuscular mycorrhizal colonization of coconut seedlings. *Proc. Indian. Acad. Sci.* (Plant Sci.). **97**: 289 – 294.

Thomson, B.D., Robson, A.D. and Abbot, L.K. 1986. Effects of phosphorus on the formation of mycorrhizas by *Gigaspora calospora* and *Glomus fasciculatum* in relation to root carbohydrate. *New Phytol.* **103**: 751 – 765.

Tilak, K.V.B. R. and Murthy, B.N. 1987. Association of VA mycorrhizal fungi with the roots of different cultivars of barley. *Curr. Sci.* **564**: 1114 – 1116.

Tilak, K.V.B.R. 1995. Vesicular – arbuscular – mycorrhiza and *Azospirillum brasilense* rhizocoenosis in pearl millet in semi arid tropics. In: Adholeya, A. and Singh, S. Mycorrhizae: *biofertilizers for the future*. Proceedings of the Third National Conference in Mycorrhiza, Tata Energy Research Institute, New Delhi, 13 – 14 March, 1995. pp. 177 – 179.

Toljander, J. 2006. Interactions between soil bacteria and arbuscular mycorrhizal fungi. Doctoral Dissertation, Department of Forest Mycology and Plant Pathology, SLU, *Acta Universitatis Agriculturae Sueciae*. Vol. 2006: 39.

Treseder, K.K. 2004. A meta-analysis of mycorrhizal response to nitrogen, phosphorus and atmospheric CO_2 in field studies *New Phytol*. **164**: 347 – 355.

Trouvelot, A., Kough, J. and Gianinazzi – Pearson, V.1986. Evaluation of VA infection levels in root systems. Research for estimation methods having a functional significance. In: Gianinazzi – Pearson, V., Gianinazzi, S. (Eds.) *Physiological and Genetical Aspects of Mycorrhizae*. Proceedings of the 1st European Symposium on Mycorrhizae. INRA, Paris. pp. 217 – 221.

Valentine, A.J., Mortimer, P.E., Lintnaar, M. and Borgo, R. 2006. Drought responses of arbuscular mycorrhizal grape vines. *Symbiosis*. **41**: 127 – 133.

Valsalakumar, N., Ray, JG. and Potty, V.P. 2007. Arbuscular mycorrhizal fungi associated with green gram in South India, *Agron J*. **99**: 1260 – 1264.

Van, L.W. and Tran, T.S. 1990. Relationship between crop response and available phosphorus by the Kelowna and EDTA and DTPA- modified multiple element extractactants. *Soil Sci*. **149**: 331-338.

Vassilev, N., Vassilev, M., Azcon, R. and Meotina, A. 2001. Preparation of gel-entrapped mycorrhizal inoculum in the presence or absence of *Yarrowia lipolytica*. *Biotechnol. Lett.* **23**: 907 – 909.

Vasuvat, Y., Wadisirisak .P., Toomsen, B., Nopamornbodi, O., Thamsurakul, S., Thananusont, V. and Boonkerd, N. 1987. Interaction between mycorrhizal fungi and cowpea rhizobia on peanut cultivar Tainan – 9 in Thailand. In: Sylvia, D.M., Hung, L.L., Graham, J.H. (Eds.) *Proc. of the Seventh North American Conference on Mycorrhiza.* University of Florida, 3-8 May, Gainesville, Florida. p 61.

Vázquez, M.M., César, S., Azcón, R. and Barea, J.M. 2000. Interactions between arbuscular mycorrhizal fungi and other microbial inoculants (*Azospirillum, Pseudomonas, Trichoderma*) and their effects on microbial population and enzyme activities in the rhizosphere of maize plants. *App Soil Ecol.* **15**: 261 – 272.

Vestberg, M. 1992. Arbuscular mycorrhizal inoculation of micropropagated strawberry and field observations in Finland. *Agronomie.* **12**: 865 – 867.

Vijayakumar, B.S. and Bhiravamurthy, P.V. 1999. Selection of the efficient indigenous local VAM fungi for improving the groundnut crop in semi-arid topical soils of Anantapur district, Andhra Pradesh. In: *Proc. Natl. Conf on Mycorrhiza.* 5 – 7 March Bhopal, India.

Vijayalakshmi, M. and Rao, A.S. 1988. Vesicular – arbuscular mycorrhizal association in sesame. *Proc. Indian. Acad. Sci.* (Plant Sci.). **98**: 55 – 59.

Vijayalakshmi, M. and Rao, A.S. 1993. Influence of fungicides on vesicular – arbuscular mycorrhiza in *Sesamum indicum* L. *Microbiol. Res.* **148**: 483 – 486.

Vosátka, M. 1995. Influence of inoculation with arbuscular mycorrhizal fungi on the growth and mycorrhizal infection of transplanted onion. *Agric. Ecosys. Environ.* **53**: 151 – 159.

Walkley, A. and Black, L.A. 1934. An examination of the Degtareff method for determining soil organic matter and a proposed modification of the chromic acid titration method. *Soil Sci.* **37**: 29 – 38.

Wang, G.M., Stribley, D.P, Tinker, P.B. and Walker, C. 1993. Effects of pH on arbuscular mycorrhiza 1. Field observations on the long-term liming experiments at Rothamstead and Woburn. *New Phytol.* **124**: 465-472.

Wantanabe, F.S. and Olsen, S.R. 1965. Test of an ascorbic acid method for determining phosphorus in water and $NaHCO_3$ extract from soil. *Soil Sci. Soc. Am. J.* **29**: 677 – 678.

Warner, A. and Mosse, B. 1980. Independent spread of vesicular arbuscular mycorrhizal fungi in Soil. *Trans. Br. Mycol. Soc.* **74**: 407 – 410.

Watson, D.I. 1958. The dependence of net assimilation rate on leaf area index. *Ann. Bot.* **22**: 37 – 54.

Watson, D.J. 1947. *The Physiological Basis of Variation in Yield.* Adv. Agron. Academic Press, Inc. New York 4th ed. pp. 101 – 145.

Weiss, E.A. 1983. *Oilseed crops* Longman, New York. pp. 282 – 340.

White, J.A. and Brown, M.F. 1979. Ultra structure and X-ray analysis of phosphorus granules in a vesicular – arbuscular mycorrhizal fungus. *Can. J. Bot.* **57**: 2812 – 2818.

Widiastuti, H. and Taharadi, J.S. 1993. Effect of vesicular arbuscular mycorrhizal inoculation on the growth and nutrient uptake of micropropagated oil palm. *Menara Perkebunan.* **61**: 56 – 60.

Williams, R.F. 1946. The physiology of the plant growth with special reference to the concept of NAR. *Ann. Bot.* **10**: 41 – 72.

Wilson, D.O. 1988. Differential plant response to inoculation with two VA mycorrhizal fungi isolated from low - pH soil. *Plant Soil.* **110**: 69-75.

Wilson, J.M. and Tommerup, I.C. 1992. Interactions between fungal symbionts: VA mycorrhizae. In: Allen, M.F. (Ed.) Mycorrhizal *Functioning: An Integrative Plant – Fungal Process.* Chapman and Hall, New York pp. 199 – 248.

Wyss, P., Mello, R.B. and Wiemken, A. 1990. Vesicular arbuscular mycorrhizas of wild-type soybean and non-nodulating mutants with *Glomus mosseae* contain symbiosis – specific polypeptides (mycorrhizins) immunologically cross-reactive with nodulins. *Planta.* **182**: 22.

Yemm, E.W. and Cocking, E.C.1955. Determination of amino acids with ninhydrin. *Analyst* **80**: 209 – 230.

Zak, M.R. 2009. Arbuscular mycorrhizae of dominant plant species in Yungan forests, Argentina. *Mycologia* **101**: 612-621.

Zar, J.H. 1999. Biostatistical analyis – Pearson Education (Singapore) Pte Ltd., Delhi, India pp. 273 – 281.

Zézé, A., Brou, Y.C., Medich, A. and Marty, F. 2001. Molecular characterization of a mycorrhizal inoculant that enhances *Trifolium alexandrium* resistance under water stress conditions. *African J. Biotechnol.* **6**: 1524 – 1528.

Zhu, Y-G., Smith, F.A. and Smith, S.E. 2003. Phosphorus efficiencies and responses of barley (*Hordeum vulgare* L.) to arbuscular mycorrhizal fungi grown in highly calcareous soils. *Mycorrhiza* **13**: 93-100.

About the Authors

R. Rajasree is Senior Teacher at the NSS Higher Secondary School, Kaviyoor, Kerala, India. After her M.Sc (Botany) from the University of Kerala, she received her PhD (Botany) in 2010 from Mahatma Gandhi University, Kottayam, Kerala. Her area of interest is plant-microbe interactions – particularly arbuscular mycorrhiza.

V.S. Harikumar is Associate Professor and Head of the Department of Post Graduate Studies & Research in Botany, Sanatana Dharma College (University of Kerala), Alappuzha, Kerala, India.

He received his PhD in 1998 from the University of Kerala in the field of Agricultural Microbiology. His major area of interest is Microbial Biotechnology. He has more than two decades of Post Graduate teaching experience and has a good number of publications in referred journals. He is a member of the editorial board of journals and member of various academic bodies. He also guides students for doctoral programmes in the fields of Botany, Microbology and Biotechnology.